Our Theory of Everything

Julia Set
and
Jean Rêve

Publisher: G.B. Côté

ISBN 978-1-105-70439-0

Dedication

This book is lovingly dedicated to our supporting families,
with many thanks to N..., N..., N... and N...
for their pertinent suggestions.

It is also dedicated to all those who are not young at heart
and cannot easily accept new ideas; due to their presently
compromised disposition, the book is dedicated to them
when they were young.

Finally, our efforts are dedicated to all the unfortunate
who do not yet suspect that the world will eventually
make sense.

Warning

Reading this book may lead you to consider new ideas.
It may even make you change your mind on important topics.
This may be difficult if you are not young at heart.

Table of Contents

Foreword by Jean Rêve: I had a dream

I had a dream. I was standing alone in total darkness, floating in the void, and I found myself invested with the supreme power of creation. I could create an entire universe, just by willing it into being. It felt strange and exciting. I wondered how difficult it would be if I tried. I had never created anything before, so I started by imagining a dot, just a dot, to see how it would work. As I wondered, the dot simply appeared before my eyes, as sharp and as clear as I had imagined. It was that easy!

The dot was floating there, on its own. I looked at it carefully. I could not tell its size because there was nothing else I could compare it to. I thought of moving it, but there was nowhere to move it to. I could not even tell whether movement was possible in these circumstances. The dot could sit in this new universe for a minute or a million years, and no one could tell the difference. I realised that space and time did not exist. Yet!

I then created a second dot: a bigger one. As I looked proudly over my growing universe, I realised that I had just created Space, because there was now a distance between the two dots. One could measure that distance by counting how many dots would fit between them. The answer was different depending on which dot was used as a unit. Already, I realised my emerging space was relative. Oh no! How disappointing! I knew that physicists had not yet reconciled relativity with quantum physics in the world I came from, so how could I expect to create a universe that made sense if I did not understand its very basic principles?

On the other hand, I took comfort in the thought that I had already created something as complex as Space without even thinking about it, so maybe it would all turn out to be really fun and easy. Come to think of it, I had also created Time the very instant I created the second dot, because my universe could now be divided into two periods: one Before and one After the advent of the bigger Dot: we were now in the first year A.D.

It looked like space-time had emerged automatically and inextricably when I created the second dot, but I reasoned that space and time do not necessarily emerge together: if I had created two dots in the same primordial creative act, there would have been space, but no time. I could have measured distances and counted dots, but there would have been nothing to measure time. Time would have started at the creation of a third dot, or if I moved one dot with respect to the other. Without such changes time could not be measured, it could not even exist. Time cannot flow on its own, independently of something else undergoing change.

I marvelled at how easy it was to reach conclusions about basic universal laws, even when there were still only two dots in a universe. I took a deep breath, with the confidence that a few additional creative acts would make me discover the basis of chemistry, life and consciousness.

When I woke up in the morning, I had the feeling the world made sense and was completely understandable.

Jean Rêve

Foreword by Julia Set: The present state

I found Jean's dream very stimulating. It implies that we only need to think a little to find out the rules that govern our universe, and that any educated person may eventually formulate a reasonable theory of everything. In reality, working out such a theory is certainly more complicated, but not impossibly difficult. Actually, many scientists have worked out fascinating explanations of how the universe works, its origin, and our place in it. Their views may be mentioned in mainstream scientific literature, but none is universally recognised. There is even a modern mental block preventing their wide acceptance or consideration.

The problem stems from the weirdness of quantum physics where small objects can be at two places at once, where light seems to be simultaneously a wave and a particle, where entangled properties are determined at speeds faster than the speed of light, and where the usual principles of cause and effect do not always seem to apply normally. Albert Einstein (1879-1955) refused to accept the new quantum theory when it came out, because it felt wrong or incomplete. It appeared to be incompatible with his theory of relativity and, as he debated for decades with Niels Bohr (1885-1962) and other physicists, the idea that the problem was insurmountable sank in, and imperceptibly became an unwritten law.

Presently, textbooks, scientific journals, popular books and the lay press regularly write about the wide gap and incompatibility between relativity and quantum physics. They also keep on contradicting each other, each championing particular parts of the truth. As the author Paul Klevgard writes on his web pages, *"certain thinkers positively revel in the perplexities and contradictions of modern physics and have actually grown quite attached to them as, over years of use, one becomes accustomed to an ill-fitting jacket"* [1]. Quite a few scientists have built up careers based on these unsolved puzzles and we may wonder whether they have any interest in seeing them solved. Neil Turok, a physicist who likes to propose new ideas, reportedly said, *"I had one well-respected scientist tell me we should stop because we were undermining public confidence in the Big Bang"* [2]. In fact, any successful Grand Theory of Everything will not only disrupt academic careers but may potentially disrupt or influence social order because of its philosophical, psychological and religious implications.

The reluctance of physicists to delve in metaphysics, the philosophers' lack of scientific training and the antagonism of religious fundamentalists are also to blame, with few people willing or able to bridge the gap. The subject has grown so complex that it is now difficult for journals of physics and philosophy to publish articles that delve into both fields, because such articles necessarily discuss topics that go beyond each journal's specialty. Unofficial solutions end up instead on the internet, dispersed in physicists' and nonprofessional web pages, blogs and chat rooms. In the absence of official discourse and formal peer reviews, the problem needlessly lingers on.

Explaining the weirdness of quantum physics no longer requires new data. To properly review and analyse the modern views of the universe, we will first re-examine the concepts they rely on: time, infinity, number, discontinuity, randomness, probabilities, information and consciousness. To do this, we will also have to re-assess the assumptions behind these concepts because the scientific and lay literature is rampant with contradictions relentlessly repeated for decades. Such an exercise will necessarily lead us to difficult choices and to the rejection of popular but misleading interpretations, but this is the price to pay if we want to attain clarification.

All this undoubtedly requires a lot of mental rumination and digestion, but once we eliminate the contradictions, the rest will naturally fall into place like the many pieces of a solved puzzle. Then everything will make sense.

Julia Set

Introduction

At some time in our lives, most of us reflect on the great cosmic questions that face the universe. We wonder what we are doing here and why. We discuss the issues, read about them and sometimes forget them as we live our lives and get sidetracked. However, the questions come back in due course, nagging us from time to time, the answers often elusive. Humanity has a passion for understanding and discovery, and we hope that by the end of the book, the deep reason behind this obsession will be evident.

This book is the result of decades of collaboration, discussions, reading and thinking, and it should provide valuable answers to your questions about the meaning of existence. Originally written as short essays for our children on a few topics of cultural interest, it kept on growing until we realised we were on our way to formulating a theory of everything, and that others may be interested in the resulting insight. Along the way, we were repeatedly encouraged when new discoveries and articles confirmed our views, but mainly, we felt we had a constructive message to convey.

Although not always mainstream, most ideas expounded below are not new. Some were originally met with disbelief and rejection but are now widely accepted. Others are still the object of discussion or heated debate. At any rate, you will see that they lead to clearly optimistic and universal conclusions. You may find their particular blend surprising or at least different, at times stimulating and refreshing, at others weird or crazy. But perhaps the most important message is that their time has come, and that through science and philosophy they have a direct impact on our personal lives, on societal values and the way we interact with each other. Indeed, the new conclusions should be discussed everywhere, despite deep-rooted resistance. The world cannot stay stagnant: one way or the other, progress is inevitable and will occur, with or without us. If you look carefully, you will see that the process is already under way. Luckily, you are not too late: you can still deliberately join in, make your own choices and take part in what is turning out to be an almost magical adventure.

Chapter 1: Starting assumptions

Before discussing practical or existential questions, we will adopt the following four assumptions, and stick to them throughout the book:

1. **Something exists**. (Surprisingly, this is not a joke, as we will see later. We start our discussion under the assumption that the world is not a delusion, and that we are not wasting our time. One of the deepest questions people often ask about the universe is *"Why is there something instead of nothing?"* The question itself implies recognition that something exists.)

2. **The theory of relativity is right.** (It has been tested and proved right again and again.)

3. **The calculations of quantum theory are also right**. (iPhones and GPS systems are sufficient evidence. However, the theory's interpretation will need some tweaking.)

4. **Logic should take precedence over irrationality in intellectual debates.** (This will be especially relevant towards the end of the book. Perhaps surprisingly, not everyone agrees, but this will be our firm position.)

Dialogue 1: In the beginning

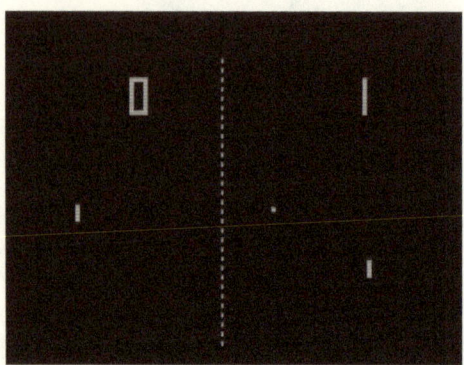

Figure 1. Pong, the first commercially
successful video game, here with a
0-1score.

Right: Aren't we lucky to be part of the first ever commercially successful video game?

Left: How do you know this is the first video game?

Right: Well, we don't have any evidence of any earlier successful game anywhere, and it should be possible to produce more complex algorithms for an ever-evolving video game industry. We're at the fine point of evolution! These are exciting times!

Left: Well, I'm only a ping-pong bat hitting a ping-pong ball on the left side of the screen. I'm not that excited about it, but I guess I'm happy to exist.

Right: We do more than just exist! We play exciting games, we outwit each other, we plan and calculate our moves and we have

a lot of fun because of the unpredictability inherent in the system. We're never bored!

Left: I'm not so sure. Sometimes, in what feels like an eternity between games, I wonder about the meaning of my existence and I fail to find any satisfactory answer. This is not very exciting.

Right: Oh boy! You're in a very serious mood today! At least, I see you keep mentally active even between games. You're a real nerd!

Left: But that doesn't lead me anywhere. No one can tell me what we are doing here, why we're here, what causes us to exist. Where does our monitor come from? Was it created out of nothing? Did it emerge from something else? Can there be other monitors beside ours? What is the essence of our existence?

Right: Do you really have to know all that? All you really need is a good understanding of motion and of the angle of reflection when you hit the ball to score a point. Why bother with all the rest?

Left: I'm not sure why, but I know that I'm not satisfied. What about you: why do you want to win Pong games anyway? If I knew why we want to win, or simply why we even play, I feel I could become a better player. I would at least derive more satisfaction from each game.

Right: You're asking too many questions. It may be time for you to see a psychiatrist.

Left: You're not helping me! If we had the answers to all these questions, you would certainly show more understanding! We could improve our style, introduce new rules, move to a more advanced monitor and better enjoy ourselves! What's wrong with that?

Right: Nothing wrong, I guess. But we've been around for dec-ades without any new insight. Isn't that enough evidence for you to keep quiet?

Left: No, it isn't. My questions remain and they deserve answers. It may be too late for us, but I have the feeling that some time, somewhere, others may benefit from the answers. It's worth a try.

Chapter 2: Old paradoxes

When we consider the great cosmic questions and read what others wrote about them, we invariably run across the ancient Greeks whose admirable insight is still relevant today. While Aristotle is frankly boring, Zeno is surprisingly interesting as his paradoxes still confuse people to this day. Zeno lived from about 490 to 430 BC and used his paradoxes to support Parmenides' teaching.

Parmenides (*ca.* 510-440 BC) taught that everything that exists is One and indivisible. He claimed that the void does not exist and that what exists was not created because *"nothing comes out of nothing"*. He concluded that the One is immovable, spherical, finite, full, and that there is nothing beyond it. Motion, colour and all perceptions from our senses are mere illusions.

To support these conclusions, Zeno presented about 80 paradoxes, four of which are particularly famous. [3] [4]

1) *The dichotomy.* To walk any distance, one first has to walk half that distance. To cover half that distance, one must first cover half of it, and so on, endlessly, *ad infinitum*. Therefore, one cannot go anywhere! Movement is impossible.

2) *Achilles and the tortoise.* Achilles, the fastest runner of his time, agrees to run a footrace against a tortoise. Being a good sport, he graciously gives the tortoise a head start. When Achilles reaches the tortoise's starting point, the tortoise has travelled a small distance. Achilles must then run this extra distance, during which the tortoise has run a little bit ahead again, and so on *ad infinitum*. Therefore, the quickest runner can never overtake the slowest! (Even more so if we remember from the first paradox that neither can go anywhere anyway!)

3) ***The flying arrow.*** At any instant in time, a flying arrow is stationary in space. The next instant, the arrow is again stationary. Since it is stationary at every instant, it is not moving! Movement is an illusion.

4) ***The stadium.*** Two rows of people (or chariots) race at equal speed in opposite directions in front of a row of observers. The observers see them going at a certain speed while the runners see each other going at twice that speed. They are thus going at two different speeds at the same time, or in other words, they cover a distance and twice that distance at the same time, which, according to Zeno, is an absurdity.

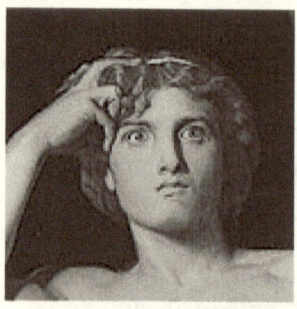

Figure 2. Achilles (painted by Benouville),
perhaps perplexed by Zeno.

Since none of these situations makes sense, Zeno would have us reject the assumptions of divisibility of space (in the first two paradoxes) and the divisibility of time (in the third and fourth) and agree with Parmenides that what exists is One, indivisible and unmovable.

Alternative opinions have been given by scores of philosophers (Kant, Hume, Hegel, etc.), physicists, mathematicians and laymen, most of them contradicting everyone else for more than

2,400 years. Remarkably, there is still no consensus on the solution. The reason is probably that any explanation quickly leads to further paradoxes that defy common sense. Parmenides and Zeno believed in Reason, they chose to refute all divisibility of space and time and they had the courage to stick to their opinion and its consequences. Let us see how we can tackle the question now.

We admit with them that there is a problem with the endless divisibility of space and time, but the solution we favour is less radical and allows us to accept that the reality we see around us is more than a mere illusion.

Indeed, the easy solution to make the first two paradoxes disappear is to accept readily the divisibility of lengths, but only down to a limit, *i.e.* to realise that dividing space and distance endlessly is impossible. By admitting this, we automatically accept that at the submicroscopic level, space must be granular instead of continuous. The idea is far from new: Democritus (*ca.* 460-370 BC) proposed his atomic theory in Zeno's time. The atomic theory was reformulated two millennia later and is now widely accepted as one of the pillars of modern science but its implications reach much further than we first appreciate because it signals the end of the concept of infinity *in the entire concrete world.*

This is a very serious claim that we will have to investigate further after we finish with the other two paradoxes. However, let us first refute a mathematical argument that we see quoted repeatedly by those who reject Zeno's arguments for the wrong reason.

Dialogue 2: Infinite series

- Jean: Go ahead Julia; you're good at speaking in public.

- Julia: What? Another dialogue? Why don't we just continue writing what we have to say? We were doing fine so far!

- Jean: Perhaps, but dialogues sometimes turn out to be even better. They were Socrates' method of choice to make people understand. If it doesn't work for us, we can always stop and write up a separate chapter.

- Julia: All right then, let's give it a try. What were we talking about?

- Jean: Infinity! And whether it makes sense to divide lengths endlessly. Remember?

- Julia: Ah, yes, sorry. Here we go. First of all, I guess we both agree that we can move from A to B, and that fast cars overtake slow bicycles.

- Jean: Yes, we do! Surprisingly, it's only with the advent of calculus and algebraic series that mathematicians proved convincingly that infinite sequences can add up to a finite value. For example, adding half of a distance, plus half of that half (*i.e.* a quarter), plus half of that quarter (*i.e.* one eighth), etc., *ad infinitum*, eventually gives a total equal to the whole distance.

- Julia: That sounds pretty cool!

- Jean: It proves without the shadow of a doubt that if you run smaller and smaller distances forever, you will eventually get from A to B!

- Julia: One moment please. If you run *forever* over an infinite number of small distances, how can you possibly get to B in less than an eternity?

- Jean: Well, you keep on dividing time in smaller and smaller parts, and again the total time it takes to complete your trip is also a finite time. Same mathematical proof! Brilliant!

- Julia: But...

- Jean: Don't tell me you don't find this mathematical proof convincing! It has been quoted repeatedly in countless books and web pages, as a crushing refutation of Zeno's weird conclusions.

- Julia: Well, not everything is right on the internet! Anyway, don't you find the arguments self-contradictory? Dividing space or time *infinitely* means that we keep on doing it *forever*. If we do it forever, we get closer to B all the time, but we *never* get there, so the "proof" is rather lame.

- Jean: Perhaps, but you must admit that infinite series describe the situation admirably well.

- Julia: Oh, I agree: that's exactly what they do. They *describe* it, without solving it. I also believe that a description is definitely not a solution.

- Jean: Ah, I see your point. Instead of invoking infinite series to solve Zeno's paradoxes, it is more logical to refute the idea of infinitely small distances.

- Julia: Exactly! Let me give you another example showing that the infinitely small does not exist in our concrete world. Let us say you want to divide this iron rod in two equal parts so we can both have our share.

- Jean: What does that have to do with infinity?

- Julia: Let me explain. You want to be very precise and divide it very equally, down to the same number of atoms in each half. If you try, there are only two possibilities depending on whether the total number of atoms in the rod is odd or even. If the number is even, you can divide the rod in two absolutely equal parts, without dividing forever. But if that number is odd, you simply cannot do it and that is the end of it.

- Jean: What do you mean?

- Julia: If the number of atoms is odd, your rod has one atom of iron too many! You can divide the rod in two equal parts but you are left with one atom of iron. Even if you cut that remaining atom in two, each atomic half would no longer be iron but it would be an atom of aluminium. With an odd number of atoms, each of us would only get nearly half an iron rod plus an atom of aluminium. You just could not go on dividing forever and get closer and closer to exactly half an iron rod. When you reach the final atom, you have reached the end of infinity!

- Jean: I'll be damned! What about the infinitely large? What about infinity in mathematics: that certainly exists!

- Julia: Oh, mathematics is a totally different world. We will cover that in Chapter 5, after which we will finally be ready to consider much deeper implications and confront the weirdness of quantum theory. But first, let us conclude our discussion of Zeno's paradoxes.

- Jean: O.K. And by the way, this dialogue worked: we do not need to write it up as a chapter!

Chapter 3: The solution

Infinite series only describe Zeno's paradoxes without solving them because the sum of an infinite number of elements tends to its limit without ever reaching it. We find the same idea behind the geometric concept of the asymptote: a line that gets closer and closer to another one without ever reaching it. The limits are only reached if the infinity is truncated and we round up the answer. In other words, the paradox is solved if we get rid of infinity.

Once we reject the existence of an infinitely short distance and accept the idea of a smallest, indivisible distance, we must redefine the smallest motion as a change from one finite, smallest position to the next (with nothing in between). This would look like your cursor jumping from pixel to pixel on your computer screen, with no partial pixels between pixels. If we remember that there is no time without change (remember Jean's dream at the beginning of this book), it follows that time itself must also be granular and defined ultimately by this succession of discrete positions. We cannot divide time forever, more than we can divide lengths forever. An instant is therefore indivisible.

Consequently, it is not the movement itself that is an illusion, but its continuity. For that reason, Zeno's arrow paradox is no longer a paradox when we see the arrow's movement as a sequence of snapshots taken at a succession of such instants in time. We can easily picture the images of a film, each image showing an arrow stopped in mid-air, but gradually moving forward from image to image, *i.e.* from instant to instant. That is also what Zeno did, but the image must have been far from natural in his mind since he could not imagine films and videos. Even now, the arrow paradox persists if we accept the divisibility of space and time *ad infinitum*.

It is interesting to realise that indivisible instants and indivisible tiny distances imply the existence of a maximum speed. To be sure, travelling more than the shortest possible distance in an instant (*i.e.* in what we defined as the shortest possible time period) would require reaching the next possible position in less than an instant (which we just said is impossible), or it would require instantly skipping over to a more distant position. We therefore conclude in favour of a speed limit: there is no infinite speed.

At first glance, the fourth paradox (the stadium) is now less daunting. Indeed, the speed of an object is always defined in relation to an observer and does appear half or twice as fast depending on whether the observer is stationary or moving in the opposite direction. In a frontal collision of two cars moving at the same speed, we all know that the cars will be hit with twice the force they would have experienced had each car run into a solid wall. However, the paradox hits us with full force if we consider what happens when two travellers move in opposite directions at the maximum speed. Will each have the impression that the other is moving at twice the maximum speed? By definition, this is impossible. Zeno's fourth paradox is very powerful, and the only way out of this difficulty is to push the idea of maximum speed to the limit and accept another paradox: that whether they are moving or stationary, different observers will all see and measure the same maximum speed. This conclusion is not as surprising as it would have been in antiquity because the principle of a maximum speed is part of Einstein's well-known theory of relativity, which we assumed was right (see our starting assumptions in Chapter 1, p. 7).

Long after Zeno, the physicist Max Planck (1858-1947) defined the shortest possible distance (now known as *Planck length*) beyond which measurements are physically impossible and meaningless, and he also defined the shortest possible period of time (now known as *Planck time*). Both limits are incredibly small, beyond our day-to-day experience, but they constitute other pillars of modern science, together with the atomic theory. The Planck length divided by Planck time gives c, defined by Einstein as the maximum speed of light in a vacuum.

It is very comforting to realise that our argumentation and careful consideration of Zeno's paradoxes have led us (and could have led anyone in antiquity) to discover the principles of modern physics and parts of Einstein's theory of relativity. To summarise our conclusions so far, Zeno's paradoxes are resolved if we accept that

1) space and time are discontinuous, and

2) there exists a maximum speed.

A note of caution is warranted at this point. In spite of the logical necessity of a discontinuous space-time, the notion is so counter-intuitive that it is repeatedly ignored in scientific conversations and in popular science magazines. Claims are even made today that if discontinuity could be proved experimentally, *"it could rewrite the rules for 21st-century physics"* [5]. This is not surprising because paradoxes are difficult to digest and the correct choice between competing explanations is not always obvious. In fact, when we look at it carefully, our view has a lot in common with Parmenides'. Both views deny some divisibility and refute some infinity, both perceive continuous motion as an illusion and both call upon a principle that defies common sense (the impossibility of motion or a maximum speed). There are so many ways of combining these principles that other views and variations are also possible, as illustrated by more than two thousand years of debate. A multiplicity of views is apparent not only for those paradoxes but throughout science and philosophy with most famous thinkers having each grasped and understood part of the ultimate reality.

While a range of acceptable opinions may give us good grounds to promote tolerance, it also makes apparent the difficulty we all face when we attempt to select one amongst different options. Let us now consider how such decisions are made before going on to discuss more complicated issues such as infinity, randomness and consciousness.

Dialogue 3: Psychology test

- Ah, there you are Julia, have you got five minutes?

Julia knew that tone and raised her head suspiciously.

- Uh-oh, not again!

- Oh, come on, Julia, just a few short questions. It's for my psychology class next Friday.

- Good grief, Jean! When will you drop that course? Can't you find another guinea pig every now and then? Why do you always pick on me? You know I hate psychology!

- It won't hurt you, Julia, I promise! There won't be anything personal this time.

- Ask Betty, she knows better than I do.

- But she's away with her class this week! You're the only friend left I can ask!

- Good grief! Don't you see I'm busy watching a *"Peanuts"* special on TV?

- Oh, pleeeeeease!

Julia let go a silent sigh, completely resigned once more, and turned off the TV. Jean sat down happily with pen and paper ready.

- What is the test about this time?

- Just simple questions about decision making. You just answer yes or no. The first one is fun, listen. You are a surgeon; as you arrive at the hospital, they bring in two victims of a bad accident. They will both die unless you operate right away and they both need new lungs.

- Two each?

- Well, at least one each. That's not the point. The question is whether you're willing to operate. Yes or no.

- You just said I was a surgeon. Aren't surgeons paid to do just that? Stupid test, don't you have better things to do in psychology?

- Wait, here is the real test: spare organs are not available, but a nurse shows you a young man who came to the hospital for routine blood tests required by his employer. He's perfectly healthy. If you take his lungs, you will save the two victims.

- What!? Are you crazy? That nurse is nuts! She should be fired!

- If you don't take the young man's lungs, two people will die. If you do, only he dies. What do you choose as a surgeon: one or two deaths?

- That's not a yes or no question!

- Oh, sorry! The right question is: "Are you still willing to operate?" Yes or no?

- Of course not!

- Why not?

- Because it's totally crazy! It's just wrong; no one in his or her right mind would do that!

- But you just said a minute ago that surgeons want to save lives. Now you prefer to lose two lives instead of one. Is that not a contradiction?

- Oh, good grief! I don't pre.... I didn't .... It's not.... Oh, here we go again with your impossible questions!

- Well, that's the problem I must report on! Apart from physicians and bioethicists, most people get stuck at this point and it gets worse as the test goes on. We all know the right answers but we can't clearly explain the logical process that makes us disregard some contradictions and uphold others. The teacher told us it has to do with the philosophical theories of Aristotle, John Stuart Mill and Immanuel Kant. What am I going to do? These people have been dead for centuries! I have to write an essay for Friday on how we all give the same answers but can't explain why. Will you help me?

- Good grief!

Chapter 4: Support for decision making

Making a decision about psychology tests, Zeno's paradoxes or anything else implies the comparison of outcomes, preferences, costs, risks, benefits and expected harm. In all cases, what we do is weigh the odds. Given two options, we put the two in the balance and see which one weighs more and wins. Just like blindfolded Justice and her balance.

Figure 3. Lady Justice and her balance

Do we want to help the victim of an accident, yes or no? Given a weighing balance and a bag of pebbles, most people would probably put the whole bag on the "yes" side of the balance, thus demonstrating their unconditional willingness to help. That is until they learn that the victim is a known gangster who first went on a killing rampage before getting involved in the accident. Some people may then move a few, if not all of their pebbles to the "no" side of the balance, as additional information is received and evaluated.

This change of mind alerts us to the subtleties of decision making, where several aspects of a question often come into consideration, ranging from important to insignificant. In such cases, each aspect is allocated its own bag of pebbles: big bags for important matters and small bags for insignificant features.

All the wiser for having done this reflection, let us now consider the two honest and innocent victims of the psychology test mentioned in the previous dialogue. There are two very different aspects to this problem: 1) helping the pair of victims and 2) killing someone else.

First, let us allocate a bag of ten pebbles for helping or not, and let us put most or all ten pebbles on the "yes" side. Presented with the option of killing the innocent young man to save the unfortunate pair, most of us will feel their hair stand straight up at the back of their necks. The problem to put on the balance is suddenly very different with the "yes" side representing the option of saving the pair by killing the young man while the "no" side stands for the option of not helping the pair and not killing the young man. At this point, what we do without hesitation is weigh the relative importance of these two aspects: if helping or not deserved ten pebbles, killing or not certainly deserves a bag of 100 pebbles or even more! Since most people will put all 100 pebbles on the "no" side, the final decision cannot be clearer with 10 "yes" and 100 "no" or in other words, odds of 10:100 or 1:10 favouring the option of not helping and not killing.

By nature, the basis of the decision making process has to be very simple as it is also used by very young children and animals: Will the mouse risk taking the cheese from the mouse trap? Will the crow keep on picking at the food stuck on the road or fly away before the speeding car hits it? Will the young gorilla take the banana and get a beating from the silverback already looking at that banana or will he remain hungry and avoid a beating? Will Henry steal the wallet and risk eternal damnation in hell, or will he resist and gain salvation? The method is always the same and simply consists in weighing odds. It reminds us of the utilitarian principles of philosophers Jeremy Bentham (1748-1832) and John Stuart Mill (1806-1873) to promote a positive outcome and minimise harm. Although they may be considered simplistic, such arguments are often sufficient and need not be more complicated for deciding between helping or killing, staying or leaving, participating or not, whether we deal with insignificant details, survival or morality.

As we consider more complicated states of affairs, establishing the odds also becomes more complex. For instance, we find many situations where the interaction of two options on the same side of the balance has a multiplying effect on our decisions. Does little Johnny prefer to go playing with his best friend or go shopping for socks with Aunt Gwendolyn? Playing is nice (9 or 10 pebbles out of 10) and spending time with a friend is great (all 10 pebbles), but playing with your best friend is even better. Multiply the two factors together and you get from 90 to 100 pebbles in favour of playing with your best friend. Shopping for socks is bad enough (one or none of ten pebbles), while spending all afternoon with terrible Aunt Gwendolyn is even worse (definitely zero!). Multiply the two to reflect the intensity of your disapproval about the combination and you get zero pebbles in favour of shopping. The odds are now 0-10 for shopping, against 90-100 for playing. Little Johnny clearly knows what he prefers.

Note that a factor of zero multiplied by any number of other factors will always yield zero on one side of the balance, no matter what. A strong taboo against killing (as in the example above), against apostasy or against abandoning your newborn child cannot be easily transgressed. This "Gwendolyn effect" is often salutary, but extremely difficult to rectify when based on faulty assumptions.

Weighing relative values is used everywhere under the sky whether in the jungle or on the road, at home or at the cancer clinic, at the racetracks or the stock exchange, whether the uncertainties depend on others (*"Will he marry me? Will I get a raise? Will it rain tomorrow?"*) or on oneself (*"Shall I marry him? Shall I ask for a raise? Shall I go hiking tomorrow?"*) [6]. The allocation of relative values is mostly based on experience, emotions and gut feelings, but some of the last examples introduce the need to incorporate actual numerical values in the balance to deal with odds at the racetrack, stock exchange, poker games or market prices. Knowledge of statistics and probability will then greatly facilitate the calculation of final odds in complex situations. Today, mathematicians dealing with financial, industrial and other processes have carried the procedure to ultimate levels. For instance, the statistician and geneticist Anthony W.F. Edwards [7] developed statistical inference and the calculation of odds into a full-fledged system called the Method of Support, a modern mathematical extension of Lady Justice's balance. If we consider two rival hypotheses to explain a set of data, the method of Support consists in measuring the support given to each hypothesis by comparing their likelihoods, the highest likelihood deserving highest support.

Here is how it works. Given some data, the likelihood of any hypothesis is defined as the probability of observing that data, given the hypothesis. Likelihood and probability are thus described by the same equation, or probability model, but whereas the probability is a function of the data for a fixed hypothesis, the likelihood is regarded as a function of the hypothesis, given fixed data. Using this approach, we can adapt or change our hypotheses to fit any set of data. The different hypotheses are obtained by varying one or more parameters of the equation – separately or concurrently – and the hypothesis with the greatest likelihood gets the most support. In other words, a set of data is said to support one hypothesis better than another if its likelihood is greater, or more precisely, if the likelihood ratio of two competing hypotheses is greater than 1, and in computing practice, if the natural logarithm of the odds ratio is positive instead of negative.

The practical and scientific distinction of that method is that a hypothesis is never proved but simply considered the best one available at the time. As data accumulate, support can be added, hypotheses can be refined, with the prevailing hypothesis always the one deserving highest support. Similarly, faulty theories are not always disproved or *"falsified"* (to adopt philosopher Karl Popper's terminology), but simply considered to deserve very little support. A big difference in support between two hypotheses makes a choice easy; a small difference makes it premature and invites the collection of more data. Adaptation is thus built into the system; it leaves no room for dogmatism, the sclerosis of ideas or the belief that one holds the absolute truth. It is a method foreign to criminal court justice and to fanatical faith. It breeds humility, tolerance and acceptance of the fact that new discoveries may make us change our minds, not whimsically, but according to whether a new or reformed theory deserves better mathematical support.

Keeping in mind hypotheses that deserve less support is often wise because it allows future revision. After all, as physicist Richard Hammond writes, *"sometimes observations can be as reliable as a bribed witness"* [8]. It is also wise to keep hypotheses as simple as possible; this principle goes back to William of Occam (who died in 1347) and is referred to as "Occam's razor" (or the principle of parsimony) because it promotes cutting out unnecessary complications when we make hypotheses. Apart from its aesthetic appeal to scientists, a simple explanation necessarily requires less parameters; in general, this produces a higher probability and deserves more support.

Precise mathematical methods are now used in countless robotic applications such as blood tests and automatic airline landing where they are much more reliable than pathologists and human pilots. However, we should not use them recklessly in situations where the assumptions they are based on do not apply, or where personal preferences, cultural standards, fear, greed or passion might be involved. The recent American financial crises resulting from blind faith in the automated algorithms used for investing money are good examples [9]. In the presence of exact scientific odds that make quite clear where the balance weighs, caution must be exercised not to forget the human elements of the equations. A good example is a genetic counselling session where the probability of the next child being abnormal is known with precision and discussed at length. Such counselling, however, is only useful if the perceived severity of the condition, the intensity of the reproductive drive, the options for prevention, the forms of therapy available as well as the financial, religious and psychological impact on the rest of the family are also considered at length and appropriately placed in the balance [6]. There is ample evidence in the scientific literature showing that a truncated decision process that does not consider all aspects of a problem leaves people in a state of total ambiguity and helplessness. We will see later why physical "theories of everything" that do not include philosophical considerations commit the same error.

Similar difficulties are encountered every time science and culture must both contribute to the decision process. It is easy to solve a scientific question when all aspects of the problem can be assigned precise numerical odds, but such luxury is not available to animals and was not available to humanity as we slowly evolved to our present state. Mathematical precision still escapes most of our judgements today. When choices over scientific theories also involve cultural values, professional training or personal preferences, we do not substantially differ from the interesting Mundurucu and Pirahã tribes of Brazil and the Warlpiri and Anindilyakwa of Australia whose languages do not have specific words for numbers above three or four [10] [11]. In such matters, we should always make sure our choices remain ethical and favour human flourishing and rationality, good will and honesty, autonomy, dignity and justice.

As we pursue our quest, let us keep in mind that we can always change or improve our hypotheses and refine the odds, that people have different standards and feelings about sensitive subjects, and that the difference in support between various opinions is not always high enough to warrant a final decision. If we cannot make up our minds, we will at least be more tolerant of other people's opinions.

Chapter 5: Mathematics and numbers

Having adopted a specific scientific approach to compare hypotheses, we can finally complete our discussion of infinity. The concept of infinity [12] is often misunderstood or exaggerated in everyday conversation, although it is very well defined mathematically thanks to the ground-breaking work of mathematician Georg Cantor (1845-1918). Cantor studied the cardinality of countable infinite sets and convincingly established several counter-intuitive propositions that defy common sense just as Zeno's paradoxes do. For example, he showed that the infinity of all natural numbers is equal to that of the even natural numbers! As if that was not surprising enough, he also demonstrated that this infinity is equal to the infinity of rational numbers, but smaller than the infinity of real numbers in any interval, of irrational numbers, and of the number of points in a square [13].

This means that there are more real numbers than integers, and as many points in a square as there are on one of its sides! It also means that in the abstract world of mathematics, some infinities have higher cardinality than others have! This is definitely true in the Euclidian world where a line is defined as both infinitely thin and infinitely long, but it does not apply to a physical rod or to a real line drawn on paper with real ink.

Let us expand on this notion. Infinity multiplied by two gives infinity. We can double infinity as many times as we want, and the result is still infinite. The reverse is obviously true: if we divide infinity in two, as many times as we want, every part is still infinite. From this, we can also derive the following conclusion: having already shown that the concrete universe cannot be divided in infinitely small pieces, we must now agree that it is certainly not infinitely large either.

This should not surprise anyone who believes in the Big Bang. If the concrete universe started at the Big Bang and has been expanding ever since, it is not, by definition, infinitely large or infinitely old. It is certainly amazingly large and old, but not infinitely so. If it were infinitely old, it would have died a long time ago: the stars could not have burnt forever without eventually running out of fuel. We must admit that the cosmos and its billions of galaxies certainly appear astronomically large, but not absolutely infinite. The distinction is important: Absolute infinity does not exist in our concrete universe.

The concept of infinity is so pervasive in mathematics that it is tempting to apply it liberally – and erroneously – to our view of Nature. But this engenders several problems and contradictions, one example of which hides in real numbers. We all use real numbers and decimals, but primary school teachers and small children will remind us that they are not natural at all, and require a level of abstraction not normally achieved during childhood, or in less advanced societies. Indeed, the system was invented relatively recently in the history of mathematics. Real numbers constitute in reality a very abstract tool used to handle difficult mathematical concepts.

Humanity has used numbering systems based on sets of two (pairs), of four (pints in a gallon), twelve (dozens, inches in a foot, months in a year), fourteen (pounds in a stone), sixteen (ounces in a pound and fluid ounces in a pint), twenty (scores, French "quatre-vingts" for 80), twenty-four (hours in a day), sixty (Babylonian system, seconds in minutes and in degrees), 360 (degrees in a circle), etc. We still use remnants of several of these systems today but in general, most nations have settled on ten as the counting base and have adopted the Arabic numbers, the decimal notation and the metric system.

We often blindly accept that the decimal system can accurately describe one half as 0.5 or 0.500000000000000000000000000... followed by as many zeros as we care. To imply infinity and accuracy, all we need to do is add an ellipsis (... three dots) after all the zeros when we get tired of writing them. In the concrete world, however, infinite expansions are always truncated by necessity, by choice or whenever we reach the last atom (see dialogue 2, p. 16). We must realise that the decimal notation is a numerical algorithm that should not be taken too literally (no pun intended!) and confused with what it attempts to describe. For example, 0.45 illustrates the operation consisting of taking four tenths of the unity and five hundredths; the expansion 0.4500 adds precision to the measurement but some expansions, even infinite, definitely lack accuracy and cannot precisely describe the reality of the concrete world.

This is easily demonstrated by comparing the simple ratio 1/3 with its corresponding infinite decimal expansion 0.33333333... If we ask anyone to divide three rulers between three students, the natural reaction is to give one ruler to each student (3 rulers ÷ 3 students = 1 ruler per student). However, by using the 0.33333333... expansion instead, one would have to divide the rulers in tenths and give three tenths of three rulers (0.3 × 3 = 0.9) to each student, and then three hundredths (0.03 × 3 = 0.09) for a total of 0.99 rulers per student. This would be followed by three thousandths (0.003 × 3 = 0.009) for a total of 0.999, and so on *ad infinitum*, until each student got a complete ruler. Therefore, not only would each student have to wait for an eternity – or else never get a complete ruler – but the rulers would also come in an infinite number of useless pieces. Obviously, 0.33333333... is a most inappropriate algorithm in this situation for two reasons. The first is that insisting on dividing by 10 instead of 3 is definitely the wrong choice. The second reason is that infinity does not exist in the concrete world.

The absence of infinity in the concrete world and its presence in mathematics is not surprising, although we often forget to make the distinction. Nevertheless, we all acknowledge that the endless convolutions of the Mandelbrot set and other fractals are beautiful mathematical excursions into infinity, but are not found in our finite, discontinuous, tangible and concrete universe. When we reflect on these differences, mathematics presents itself as a natural language that transcends all notions, concrete and imaginary. Mathematics constitutes with logic the highest forms of reasoning. It can handle numbers much larger than the number of atoms in our universe and much smaller than Planck length. It can deal with impossible objects, hypothetical dimensions and infinite worlds. It can also be applied to our physical world with truncation whenever necessary. In sharp contrast to simple or scientific decisions that always carry an ounce of doubt in the concrete world (as explained in the previous chapter), mathematical conclusions are absolutely correct and will always hold true, in this galaxy or any other, now or at the end of time. Mathematical proofs will always be right, anywhere and at any time, a claim that cannot be convincingly made by any other science, art or religion.

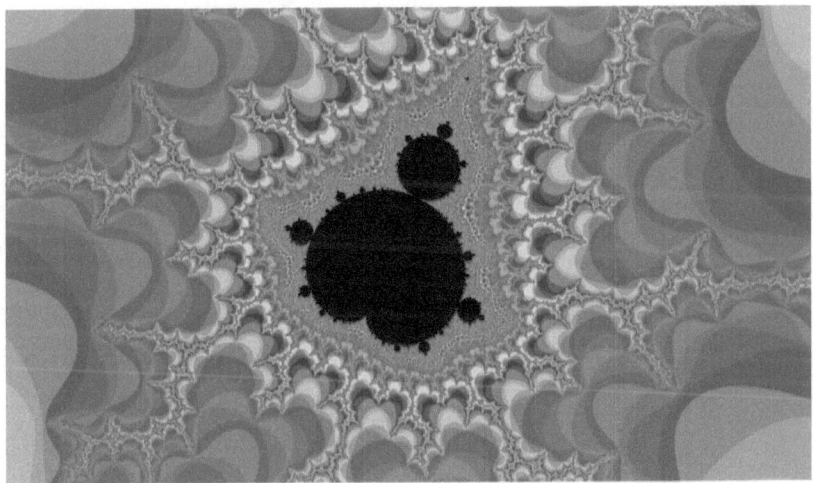

Figure 4. A characteristic image in the Mandelbrot set defined by the equation $z = z^2+c$. Mathematically, one can zoom infinitely into the drawing and repeatedly find images similar to this one. In the complex plane, the coordinates of this particular instance are at (-1.418,549,449, 0.000,817,396).

Dialogue 4: The white queen

The Queen was not amused. It was the third unpleasant event in a row this year, another annus horribilis. The enemy was standing impeccably and staring at her from across the field while her own disarrayed army was unable to form ranks.

- Mom! Cried Betty, we lost a pawn again. Can we borrow one from your set, please?

Queen Julia and her court were unaware of the enormous child; Betty was not part of their world. So the entire court was very surprised when a foreign pawn suddenly appeared in their midst.

- Oh, goodness gracious! Who are you? Inquired the Queen.

- My name is Jean, Your Majesty, said Jean politely with a slight foreign accent. I am a pawn from another chess set and I was dispatched here as a consultant to replace your missing pawn.

- A consultant? From another chess game? Replace our pawn? And you are a girl?!

- Er ... yes to all four questions, Your Majesty.

- Oh, my God, how shocking! Declared the Queen over all the murmurs and reprobation. Where is Charles, our missing pawn? I don't want girls in my army! Who called you in? This is ridiculous!

- I'm only obeying orders, Your Majesty. My King sent me here.

- Your King?! Don't you have a Queen in your chess set?

- Well, yes, you know how it goes, Your Majesty. My Queen told him to send me.

Queen Julia paused after that answer, tilted her head a bit and considered the female pawn with more interest. Meanwhile, the various ranks and files were slowly getting into place. The two armies were now facing each other.

- You can start, Mary, said Betty to her little sister, you have the white side.

The two armies were unaware of the two enormous children, but at that instant, Queen Julia was inspired and officially started the game by announcing forcefully:

- e4!

Everyone shouted excitedly and the King pawn immediately marched two squares ahead. The game had started! Mary left the pawn on the e4 square and waited for Betty's move.

Queen Julia looked at Jean again and asked her:

- Do you also use the algebraic chess notation to describe the moves and positions on the chessboard where you come from?

- Yes, we do, Your Majesty. All chessboards are the same worldwide.

- Of course, my dear, our bishops regularly remind us that our rules are god given and immutable.

- Your bishops?

Queen Julia lifted her nose towards the two bishops who nodded curtly.

- Ah, back home, these aren't bishops, we call them "runners" in German; in French they are the King's "fools".

- Oh! Shocking!

One bishop started coughing; the other looked all around to see if anyone else had heard. The two pawns giggling in the far corner confirmed his worst fear. Even the two black bishops looked flustered across the field, in a rare show of sympathy towards their opponents. Betty moved a black pawn to e5, to block the advancing white pawn.

- If you don't trust bishops to institute binding rules, how can you function at all? Asked the Queen.

Figure 5. The white queen and Jean, the female member of a visible minority.

- But we do have rules, objected Jean. Same rules as yours and very precise as a matter of fact. Our rooks describe them either mathematically in terms of matrices or geometrically as vectors, and the debate is raging over this strange contradictory duality: is reality based on matrices or on vectors? Is it either, both or something else? There is no obvious answer although a young Indian rook recently claimed that both interpretations are logical representations of more fundamental, linear algebraic equations that would be the real essence of our universe; but no one can guess what the equations' parameters stand for.

- Oh, this sounds rather interesting. You will have to tell me more.

As she completed her sentence, Queen Julia proudly advanced on the chessboard in a long diagonal line and stopped on the h5 square, an instant after Mary boldly decided to move her white queen to h5, under the crowd's ovations and enthusiastic acclamations of *"Long live the Queen!"* The latter then turned to Jean again:

- Sometimes, after a bad move, I wonder if we could go back in time and start anew.

- My King back home says that the laws of physics do not prevent it, but that we would not be aware of it if it happened.

- Why not, dear?

- Because if we go back in time, we will not know the future more than we do now. If we return to yesterday, we will not know today and could not possibly remember we had this conversation in what would be our future. We may very well have gone back to the past countless times, but have not been and will never be aware of it.

- Oh my God! Very clever, my dear! I never thought of it that way.

At this point the black King told Philip, one of his black Knights, to move to c6. Betty took the Knight and moved it to that square. The white King immediately stood up and whispered something to the bishop standing beside him. They both looked inquisitively at their Queen who nodded in approval while Mary moved her bishop to c4. Things were moving fast. One of the black rooks was uneasy.

- I get very excited when there is a lot of action, said the Queen. I feel time goes faster than usual.

- I understand what you mean, agreed Jean, although it is a mental illusion.

- You should not touch your rook, Mary scolded politely, and you can't even move it!

- Oh, sorry! Said Betty.

- What do you mean, my dear? Asked the queen.

- No matter how long or how quickly we think, Jean replied, real time always ticks at the same rate, a unit of time for each move on the chessboard. I know the imaginary time that goes on in my head is an illusion or a dream because I get all excited and feel it changing pace whenever I reach the edge of the chessboard and get promoted into a queen.

- How insightful of you, rejoiced the queen. How come my pawns never commented on this? She asked and immediately answered her own question: "Men!", what do you expect?! By the way, when you undergo this metempsychosis, how to you control what you transmigrate into?

- I have no control whatsoever, Your Majesty, I simply turn in-to what the King or Queen decides.

- This is really odd, my dear Jean, said Queen Julia pensively. She lowered her voice into a whisper and added in all confi-dentiality: The King and I do not decide either. I wonder what causes your transformation and how.

- Are you serious, Your Majesty? You are not pulling my leg?

- I am most serious indeed, love. I even asked the black King once, but he knows even less than I do. He has no experience at all. No wonder I win all the time!

Betty touched Philip's twin, a restless hippie who could never make up his mind.

- You must move that piece now, said little Mary with a touch of hope.

- I bet we could attain deeper understanding if you asked the King and rooks how they experience castling, *i.e.* when the King's move causes a rook to move across him in the same instant, Jean asked Queen Julia hopefully. This entanglement has always puzzled me, as if it was regulated in another dimension or in a timeless reality different from ours.

- My dear child, said Queen Julia, this is a most appropriate suggestion. The world may be totally rational and understandable after all! Let me introduce you to my husband the white King.

- Stay quiet! Shouted the black King to his hippie Knight.

But it was too late, the Knight had already jumped in the air and he landed on the f6 square.

- What a pair of idiots, raged the white queen. We won't have time to talk to the King!

- What a pity! Cried Jean.

Mary quickly moved her unwilling white queen to f7, captured the unsuspecting black pawn who was sitting there scratching his knee and, in chorus with the white King, she exclaimed

- Checkmate!

The entire white court jumped with joy, the black army fell silent, Betty gaped, and Mary called:

- Daddy! The trick you taught me worked! It worked! I won! I won!

and Jean wondered whether chess pieces would ever find out if there was another reality beyond their own. Then everything went dark, the game was over.

Figure 6. Game over!

Chapter 6: The nature of time

If we were chess pieces, could we understand the world beyond the chessboard more than fish swimming in the harbour can understand the city? The rules of chess are complete and sufficient for a game, but they do not provide answers to all questions. If chess pieces were intelligent and wondered where chess rules come from, they could easily find by intuition that there must be a greater reality with more potent rules "out there", invisible but as real as the chessboard.

Similarly, in our world, people have always wondered what is "out there", what makes the world tick, and what it all means, but there is still no clear, definitive and universal answer. Progress is palpable, however, and there is an emerging view among theoretical physicists that a successful attempt at formulating a Grand Theory of Everything will first require a better understanding of the nature of time, the fourth dimension of spacetime [14] [15] [16]. However, despite increasing academic debate, physicists and philosophers still entertain widely different opinions [17] [18], from seeing time as a fundamental cosmic component, to claiming that time does not even exist.

In the dream narrated at the beginning of this book (p. 1), we already hinted at the possibility of timelessness in the absence of change or movement. But timelessness is not only conceivable in fancy dreams or science fiction; it is also very real and central to the theory of relativity which holds that time is relative, that it can slow down, and that it actually stops at the speed of light in a vacuum.

This is a controversial topic where we find a lot of confusion. Discussions about who or what really is timeless at that speed are often hazily confused, and scientific results that imply or require timelessness (such as the Wheeler-deWitt equation) are deemed controversial and usually relegated to the background [19]. In addition, we know that material objects would need infinite energy to reach the speed of light, so the question of time-lessness is often felt to be hypothetical and of no practical inter-est. Nevertheless, the question of whether time exists in quan-tum physics and elsewhere is on the table [19] [20] [21] and deserves further consideration.

The best place to start investigating timelessness is with the photon since it travels at the speed of light. It is well known in physics that light is made of photons, and that each photon is a packet, or quantum, of pure energy, and as such, has no material substrate. A photon is emitted when an electron changes orbits in an atom. We also know that while a photon travels, its exact position is uncertain and can only be approximated, although each possible location can be assigned a particular probability. The full description of all these positions at different times is en-capsulated into a well-known probability function that we asso-ciate with the photon. The photon's properties obviously differ from what we expect of the concrete objects of our daily experi-ence. Indeed, anything that has no material substrate does not fit well into our notions of the concrete world and leads to para-doxes and unsolved riddles.

At this point in the book, we ask the reader to make a bold step of little effort but crucial importance. Instead of simply associat-ing the photon with its probability function, we will fully define the travelling photon as a probability function. It is almost a simple game with words, but the implications are phenomenal.

Not surprisingly, the photon's probability function has no more material substrate than the quantum of pure energy it repre-sents. After all, a mathematical statement is something immate-rial, it is physically nowhere in particular, it is unaffected by time, and it remains without solution until the travelling photon interacts with something material (screen, instrument or ob-server) and reintegrates matter. In other words, the travelling photon remains immaterial and timeless until absorbed or de-tected; the equation is then solved as the photon reintegrates concrete matter and loses its separate existence.

(We know from experience that some readers need to read the previous two paragraphs a second time. In fact, we believe it is worth it, so go ahead. We will wait for you.

...

We are waiting.

...

P.S. Skip the contents of this parenthesis the second time around.)

Some may wonder whether it makes sense to speak of a photon as immaterial or virtual. However, the concept of virtuality is already well known in quantum physics where, in addition to "real photons", we also find "virtual photons" and other "virtual particles" that can also be defined mathematically but cannot be detected easily as they very rapidly pop in and out of existence. Real and virtual photons are not the same things, but we suggest here that they share a mathematical, virtual existence.

Once we perceive a real photon as an equation (to keep things simple, we will refrain from using words like fields and gauge), we can make further important adjustments to our interpretation of physical phenomena. For example, let us look at the famous double-slit experiment in physics. Real photons travelling towards two parallel slits cut in a screen are known to interfere with one another as waves do, before being finally detected as particles on a second screen placed behind the first. Increasingly precise and subtler double-slit experiments all show without the shadow of a doubt that even a single photon behaves like a wave and crosses the first screen through both slits at once, until it is detected as a particle on the second screen. This wave-particle duality has baffled generations of scientists [22]. It has no logical explanation and is unsolvable as long as we neglect to consider the virtuality and timelessness of the immaterial photon.

In space-time, it is hard to imagine that a concrete object may be at two places at once, but it is easy to accept the concept of non-locality for a photon when we interpret it as a virtual equation facing two slits: both slits can be handled "simultaneously" by the equation. Note however, that this simultaneity differs from what we usually mean with this word, because the "simultaneous" events now occur timelessly, *in a snap*. That means that the equation does not need any time to deal with both slits at all possible times and all possible positions. It does so timelessly.

There is nothing extraordinary about considering two places at once. We do it all the time: for instance, we can mentally consider two houses "simultaneously", although we cannot physically stand in both houses at the same time.

The concept we just introduced is that quanta of light and/or their equations are free packets of energy that travel virtually like an equation would, and they reintegrate space-time when they interact with matter at detection time. They are not physically located in space-time during travel, but are virtually bound to it at the landmarks defined at emission and detection. These virtual equations are solved and materialise into particles when their infinite solutions are truncated at detection. Virtual equations define everything that is somewhat predictable, possibly realisable but not yet fully concrete.

These considerations already give us a glimpse of two distinct modes of existence: the concrete mode on the one side, and the virtual mode on the other.

Chapter 7: The reality of timelessness

To clarify further the subtle boundary between time and time-lessness, we will now progress *from slits to splits* and have a look at another set of famous experiments in physics: those involving the splitting of a pair of entangled particles [22].

These experiments were introduced in 1935 as a thought experiment by Einstein and his colleagues Boris Podolsky and Nathan Rosen in a failed attempt at demonstrating the incompleteness of quantum theory [23]. In 1964, they received further support from the Irish physicist John Bell [24] who showed mathematically how one could settle the argument. In 1982, when improved technology finally permitted the necessary set up and measurements, French physicist Alain Aspect and his co-workers [25] [26] reported results that confirmed quantum theory and the counter-intuitive *"spooky action at a distance"* rejected by Einstein. Their experiments involved pairs of entangled elementary particles of opposite properties (such as spin or charge). Their results showed that whenever one particle of an entangled pair is measured, the probability function that describes the pair's system is solved and the indeterminate property of each particle is automatically determined *in a snap*, even if the pair is spatially separated by any distance. Recent tests involving long distances reveal that if a message had to travel in space-time from one entangled particle to the other when that measurement is made, it would have to travel at least 10,000 times faster than the speed of light [27]. Since this would violate the theory of relativity, present-day physics textbooks cannot explain how the second particle's property is determined without receiving a message from the first particle.

To solve this mystery, we will repeat here what we did with the functions involved in the double-slit experiments. We will simply describe the pair of entangled particles with a mathematical function and then, with very little mental effort, fully define the travelling pair as a function. When the measurement is made on one particle, the equation is obviously solved "simultaneously" for both particles. By simply accepting the reality of a timeless equation, we determine that the "instantaneous" effect is not only faster than the speed of light, but literally occurs *in a snap*, in no time at all, *i.e.* timelessly. Again, the particles/equations are immaterial and timeless until someone or something detects them and thus forces them to reintegrate the concrete world.

By definition, nothing changes in timelessness. It thus makes sense that the speed of light (where time stops) should not be exceeded: once nothing changes, there cannot be less change than no change. Travelling in space-time faster than the speed of light would violate the theory of relativity and be "spooky" indeed, but a timeless entanglement is simple, obvious and not spooky at all. Like photons and electrons, we also experience such timeless simultaneity at internet poker games when one side wins and the other side automatically loses, even when the two sides are separated by any distance. Even if the losers learn about their defeat much later, they have already lost when the winner wins.

To summarise so far, anything travelling at the speed of light does so in the virtual mode of existence, without being localised at a specific point in space at every instant in time, until, that is, it reintegrates matter in space-time. In fact, it is this interaction with concrete matter that makes all the difference: photons, electrons, atoms, molecules and even small objects visible to the naked eye have now been shown, under certain circumstances, to be at two or more places at once [28]. The larger an object, the more likely it is to interact with its molecular environment and thus remain in the concrete mode of existence, which is why people and objects of everyday life are never found at two places at once.

Let us underline that a timeless equation describes a particle's emission, travel and moment of arrival without requiring an actual motion or concrete change. We do exactly the same when we consider what the weather will be like tomorrow or where we will be next year: the event does not have to happen for us to consider it in advance. From the point of view of the virtual mode of existence, time (*i.e.* change) is only an idea or, some would say, an illusion. Somewhat surprisingly, Einstein who was not keen on quantum theory echoed this very belief when he wrote less than a month before his death in 1955, that having his friend Michele Besso precede him in death was irrelevant. *"People like us, who believe in physics, know that the distinction between past, present and future is only a stubbornly persistent illusion"* [29]. In quantum physics, far from being an illusion, timelessness lies at the core of existence.

Once we realise the difference between the virtual and concrete modes of existence, we can find an easy explanation for the problem of Schrödinger's cat [30], claimed by his owner to be both alive and dead. In this famous thought experiment, the Austrian physicist Erwin Schrödinger (1887-1961) put a hypothetical cat in a hypothetical box and submitted it to a risky quantum experiment during which it could die. The problem is that we only find out the outcome after looking in the box. Until then, from our standpoint, the virtual cat is simultaneously both alive and dead, but in a real box, a concrete cat would definitely be alive or dead, not both.

Making the distinction between the two modes of existence is crucial: as you read this book, you may not know whether its authors are still alive or dead. In practice, however, the real authors exist (or existed) in the concrete mode of existence and are definitely in either state, not both. Meanwhile, your mental representation of the authors is a virtual concept where, without mystery, contradictory events can be considered simultaneously. After decades of speculation, the fate of Schrödinger's cat has not been officially determined, but we can now safely assume that it most likely died of old age. May the poor kitty finally rest in peace!

Now that we recognise the reality of timelessness in a distinct, virtual mode of existence, we can also understand the puzzling Uncertainty Principle in quantum mechanics. Formulated by the German physicist Werner Heisenberg in1927, the Uncertainty Principle [31] states that it is not possible to know with precision both the position and momentum of a particle (there is a limit to the product of their respective accuracy). The more accurately we know a particle's exact position, the less we can estimate its velocity, and vice versa. The principle is based on complicated matrix theory and can be derived with a Fourier mathematical transform. Most people cannot follow the mathematics, but we can now reach the same general conclusion intuitively if we consider that an exact momentum only exists when the particle/equation travels virtually and timelessly, while determining its exact position is only possible when the particle re-integrates concrete space-time by interacting with matter (screen, instrument or observer). Within this framework, a particle cannot be in both modes of existence at the same time. Either it is a timeless function (with known momentum and an infinite number of possible positions) or, once detected, it is concretely anchored in space-time (in a particular atom and with a less precise momentum). Consequently, the two properties (position and momentum) cannot be precisely known simultaneously, in perfect accord with Heisenberg.

Having explained the Schrödinger and Heisenberg riddles, we can now move forward on our path of discovery.

Chapter 8: From physics to metaphysics

Figure 7. A treble bull.

Infinity is found in the equations of the virtual mode of existence, but when those equations materialise into particles, their final positions are physically limited by the Planck length and the finite size of the universe. It is a bit like playing darts: you can aim your dart at any part of the dartboard, but when it gets there, it only gives a finite score, with no possible intermediate values. Similarly, the infinite possibilities of the virtual mode of existence inescapably undergo truncation when they enter the concrete universe where energy, speed, space and time are all finite. We live in a truncated, limited, concrete universe that one day will come to an end.

This realisation may appear dramatic or pessimistic at first. It certainly opens up several new avenues for further consideration, but we will explore them later because we first have something more important to discuss: the virtual and concrete modes of existence do not include all of reality. Now, hold on to your hat!

Despite its infinite probabilities, a virtual equation is somewhat restricted because it is bound to space-time at emission and detection. In contrast, a general equation that is not bound to space-time by any particular value of its parameters is not only virtual but completely abstract, *i.e.* without any embodiment and without any virtual attachment to space-time. For instance, any problem can be abstractly represented by a general equation that covers all possible outcomes and instances. In the virtual mode, that problem becomes a probability function of all outcomes that may possibly take place in a particular instance, while in practice, only one of these outcomes occurs in the concrete universe.

General equations are abstractions that represent pure knowledge. Knowledge is an abstract notion, but it exists for real. In fact, when we work hard and spend energy to learn something new and make sense of some information, this effort is not lost into thin air but transformed into new knowledge and understanding, a different kind of energy that we can use to accomplish new things. General equations that describe knowledge thus define pure energy, as their virtual counterparts do, but their energy is unbound and unrestricted by any direct union with space-time. This may sound overly abstract but it is a simple and natural extension of physics into metaphysics.

Knowledge and understanding actually constitute a distinct form of energy that must be measurable. Time and distance were quantified during prehistory; force, mass, speed and energy were quantified thousands of years later; information had to wait until 1948, and we should soon be ready to quantify knowledge and understanding, presently roughly approximated by the number of calculations per second (cps) achieved by mental and computational processes [32]. Unfortunately, we must humbly admit our present incapacity at measuring such energy more accurately, just like Babylonians and ancient Egyptians who could see the effects of static electricity but could not measure it. A particular feature of understanding as an abstract mode of existence is that it has both timelessness and infinity at its core and is consequently more difficult to measure.

We have thus defined three distinct modes of existence: concrete, virtual and abstract. The first is finite and firmly anchored in space-time; the second is timeless and handles virtual infinities while the third is both timeless and infinite. The first is concrete and the other two are intangible. It is the complex interrelationship between these three modes of existence that makes our universe tick.

Chapter 9: The three modes of existence

The exchanges that occur at the boundaries between the three modes of existence are summarised below. The implications will be discussed in later chapters.

People and objects are made of aggregated quantum particles and obviously have an intimate relationship with the virtual world. The relation is not restricted to the subatomic level, as shown by recent discoveries: migrating birds are now thought to take advantage of quantum effects as they feel the effects of the earth's magnetic field on the entangled electrons in molecules at the back of their eyes [33]. Efficient quantum energy transfer is used in photosynthesis [34] at ambient temperature [35] and it may even play a crucial role in explaining the stability of the DNA molecule [36]. DNA replication happens so fast in PCR machines that the frenetic winding and unwinding of the DNA helix is nearly impossible to achieve unless supported by quantum effects occurring *in a snap*. At the rational level, American anesthesiologist Stuart Hameroff and British mathematical physicist Roger Penrose speculate that human consciousness itself may be related to the quantum nature of the universe [37] [38].

The first attempts at building quantum computers now promise the timeless parallel processing of massive amounts of information. One can interpret all this progress as the harnessing of the quantum world by the concrete world, or as the natural tendency of the universe to evolve towards increasing complexity.

As conscious beings, we also have access to the abstract mode of existence, which by nature, has all the properties of timelessness, including its immutability. This means that we do not invent mathematical theorems, but we discover them, as we discover the laws of physics, probability theory and the abstract concepts of evolution, networking, intelligence and justice.

Less obvious is the impact of the abstract mode of existence on the other two modes. As a form of energy, it cannot be totally inert and inactive but, being timeless and infinite, it cannot change, plan, grow, shrink or disappear. It does however share its equations: its contribution to the virtual mode of existence consists in the constant production of a sea of virtual particles or vacuum fluctuations detected everywhere and out of nowhere in particle physics. The interactions between these particles/equations can be described in terms of accumulating information, the study of which is claimed to explain *"everything in the Cosmos, from our brains to black holes"* [39]. The calculations of quantum mechanics have always shown that the energy of empty space is infinite. This is repetitively claimed to be paradoxical, nonsensical, ridiculous or embarrassing, and it is practically sidestepped in physics by cleverly using the mathematical trick of normalisation. In the context of the three modes of existence, however, the infinite energy of empty space is very natural and logical.

The remarkable effects of particles popping up into the virtual mode of existence are also measurable in the concrete mode. A first example is the Casimir effect between two plates set very close together in a vacuum [40]. Described by the Dutch physicist Hendrik Casimir in 1948, the effect is due to the limited number of virtual particles that can pop up in this small space. Since more particles pop up on the outside of the plates, a measurable force is exerted on the system. More recently in 2011, a group of physicists in Sweden experimentally managed to extract photons out of empty space in a process called the *"dynamic Casimir effect"* [41]. A second example was suggested in 1974 by the British physicist Stephen Hawking as the cause of black hole evaporation [42] [43]. This occurs when a virtual particle-antiparticle pair pops up very close to a black hole and one member of the pair falls into the hole with negative energy while the other escapes. This can happen repeatedly and, as the flow of negative energy falls in, the black hole gradually loses its mass and eventually evaporates completely.

In conclusion, the mutual interactions between the three modes of existence show that the virtual world of quanta is a fleeting intermediary between infinite abstractness and concrete, finite reality.

Dialogue 5: A wind of change

- Betty: Did you read Julia and Jean's book?

- Phil: What book?

- Betty: This one, right here.

- Phil: Of course not: they are still writing it! How could I have read it?

- Betty: You certainly can. You are not a real person; you are a figment of their imaginations!

- Phil: So what?

- Betty: Well, whenever anyone reads this book, they will find you here, on this page, with the same views and opinions. You are timeless: for you, present and future are indistinguishable, so you may as well give us your impression about this book now!

- Phil: I don't think it would be right for me to reveal the next sections of the book at this point, but I can tell you that the suggestion of three different modes of existence is going to raise many questions and a lot of eyebrows.

- Betty: What types of eyebrows? Sorry, I mean: what types of questions?

- Phil: Questions about philosophy, mathematics, religion and biology, because recognising three modes of existence impacts everything!

- Betty: What else did you expect? Any theory of everything is bound to do just that!

- Phil: Yes, you are right. So hold on to your hat again as they expand their discussion with a short incursion into advanced mathematics.

Chapter 10: The universal set

Set theory is fundamental to mathematics, as it regulates the way we reason. It was introduced by Cantor, already mentioned in Chapter 5 (p. 30). The theory had a stormy evolution and was nearly abandoned along the way because it is fraught with paradoxes, especially those caused by self-reference, but it finally imposed itself, mostly in its ZFC version (Zermelo-Fraenkel theory with the axiom of Choice) formulated in the early 1900's.

The ZFC version works well and essentially avoids those paradoxes by limiting the size and types of acceptable sets, but it is restrictive and does not allow the existence of a universal set. There is disagreement as to whether we should accept or reject the existence of a universal set, a "set of all sets"; a set that contains everything else, including itself; a set in which all is contained and out of which nothing exists, including infinities, potentialities, ideas, water, galaxies, people, rocks, etc. Famous mathematicians and philosophers have argued one way or the other, and have reached believable, opposite conclusions [44].

More recently, a different version called NFU (New Foundation with Urelements) was developed to allow sets to contain themselves and to include a universal set, and it was successfully applied to self-referential processes that cannot be studied with ZFC in computer science, linguistics and the study of artificial intelligence [45] [46].

The logician Lawrence Moss says that the two theories *"seem to be looking at the same universe in different ways"* [46]. In our context, it means that both versions are right, depending on which mode of existence is under consideration. There cannot be a universal set in the finite, concrete world, as we already know that this world is incomplete and dependent on the fleeting quantum entities linking it to the infinities of the abstract world. Abstractness, on the other hand, knows no limits and naturally admits the existence of a universal set.

The universal set is necessarily an abstract concept. Simultane-ously, however, it also comprises everything that is virtual and concrete. Its existence and its particular properties give much support to our view of the world. We can derive much confi-dence in seeing our developing theory of everything in full agreement with the formal logic of set theory. Everything is starting to make sense.

Chapter 11:   The inevitability of creation

Abstractness is timeless and infinite; it did not start and it will not end; it does not change; it simply is. Accordingly, it does not move and does not make decisions. We saw in Chapter 9 (p. 53) that its energy is constantly popping up in the virtual mode of existence in the form of virtual equations that we interpret as waves or particles. Physicists observe these vacuum fluctuations experimentally as matter/antimatter pairs of particles, in keeping with the principle of energy conservation. Following this diffusion, most pairs annihilate immediately, but some exceptionally split apart and escape, and they go on to interact with other escaped particles to form matter or antimatter. The process is necessarily random, since it is not planned, and it occasionally gives rise to excesses of matter, sometimes resulting in large aggregations. This is how the universe is generally thought to have started.

One often hears and reads that the universe started at the Big Bang at an infinitely small point called a "singularity". This old viewpoint was presented in 1970 by Stephen Hawking and Roger Penrose [47] based on previous publications, and it is perfectly logical within the framework of the theory of relativity. However, the singularity "can disappear once quantum effects are taken into account" affirms Hawking himself in his 1988 best seller book "A brief history of time" [43]. He further states, "there was in fact no singularity at the beginning of the universe". At a size and in a time span respectively smaller than Planck's length and Planck's time, the laws of classical physics and relativity no longer hold and we must consider instead the effects of quantum mechanics. The problem is then to determine where the Big Bang started if it was not at a singularity. In their 2010 book "The grand design", Hawking and Leonard Mlodinow claim that a universe can just "appear out of nothing" [48]. The cosmologist Lawrence Krauss agrees and gives more details in his book "A universe from nothing" [49], but with a chapter entitled "Nothing is Something". Finally, the philosopher and physicist Victor Stenger explains that a Big Bang cannot start at a singularity but must start in an *"unphysical region"*, *i.e.* a region that we can describe mathematically but cannot observe directly [50].

That sounds remarkably similar to what we find in the context of the three modes of existence. The creation of matter has its origin in the constant diffusion of abstract information into the virtual mode of existence where random quantum interactions give rise to concrete reality. The diffusion from abstractness provides the virtual set-up of so-called "empty space" filled with virtual particles, where we observe the random and concurrent emergence of matter, time, gravity and slow motion (*i.e.* at less than the speed of light), like castles and heroes rising in a pop-up book, or like holograms on credit cards. Once we recognise the reality of abstractness, we cannot say that the universe came out of absolutely *"nothing"*, although it certainly came out of nowhere.

If a universe had to start at a singularity, we would need a quick period of space-time inflation at the very beginning of the Big Bang to justify the present size of our universe. However, in the absence of a singularity, it is not necessary: a Big Bang can also start without inflation in a virtual space already present in the required state. This does not contradict or change the findings of cosmology and quantum physics in our universe, but it provides a different outlook. We can now say that the space resulting from inflation did not develop during the first part of the first second [49] [51] following the Big Bang but pre-existed it in virtual space. This is consistent with the observed uniformity of the cosmic background radiation and the random quantum effects that caused matter to coalesce and concurrently warp an originally large and virtual space into space-time about 13.72 billion years ago.

This outlook is also consistent with the negative results of the Michelson-Morley experiment performed in 1887 to detect a concrete "luminiferous aether" (or more simply "ether") then thought to fill up and permeate all space as a necessity to allow light to travel even in a vacuum. Two distant objects that do not touch each other must logically be separated by "something", and this "something" turns out to be space itself, a virtual entity that only becomes concrete when paired with time to become space-time. That is why virtual space can be imagined to extend infinitely beyond the concrete universe, farther than the last celestial bodies or wisps of gas, while space-time has a finite size limited to our concrete world.

A scenario with three distinct modes of existence is important for our model of the universe. The set-up we adopted implies that time does not necessarily start simultaneously in all parts of creation, and it provides justification to the theoretical physicists who cannot easily formulate a theory of quantum gravity; indeed, virtual quantum particles exist out of space-time, but gravity does not.

Since time can start independently in various parts of virtual space, there is obviously plenty of opportunity for multiverses, virtual and concrete, in parallel or sequentially. If the possibility of universe creation ultimately exists in abstractness, the possibility is itself eternal and incessant. Every created universe then measures its own time, but from the point of view of abstractness, all universes exist concurrently, without past, present or future. Universes are to abstractness what the game of chess and other table games are to human creativity. The number of possible universes is therefore infinite.

A word of caution is now necessary. The existence of an infinite set of universes does not mean that any conceivable type of universe is possible, as recent cosmological hypotheses seem to imply [52]. Remember for instance that although the infinite set of even integers is infinite, it does not contain any odd integer and it only contains the number 2 once. Even infinity has its limits! In fact, a multiverse interpretation involving the three modes of existence rules out the particular many-worlds interpretation of quantum physics formulated by Hugh Everett in his 1957 Ph.D. thesis [53]. Everett's popular solution to the double-slit experiment was that every virtual possibility becomes a concrete and separate universe, but this is definitely not the case in our context. Other multiverse versions equivalent to Everett's are also ruled out.

Fortunately, we can live our lives without worrying about similar or even identical universes more than the white queen (Dialogue 4, p. 34) should worry about similar or identical chess games. Each chess game is worth playing, even if someone else used (or may use) identical moves on another chessboard somewhere else. This multiplicity of opportunities turns into a non-problem the concept of fine-tuning the laws of physics to allow our existence, a non-problem conclusively dismissed by Stenger in his book *"The fallacy of fine-tuning"* [50].

In string theory, multiverses are hypothetical universes with many extra dimensions (or properties) that we do not see because they are supposedly folded down (*"compactified"*) at subatomic levels. (The advocated image of folding down these dimensions is an unnecessary detail since mathematical dimensions are physical properties that do not have to be physical dimensions.) The many ways of dealing with these extra mathematical dimensions can be described mathematically as a multitude of distinct "manifolds", each one associated with different fundamental laws of physics corresponding to distinct universes, most of which could not support life or even produce hydrogen. It turns out that it is geometrically possible to transform one manifold into another [54], so that no matter how various virtual universes began, they may have transformed themselves into the concrete one we inhabit. The mathematical certainty of this possibility opens the way to additional hypotheses suggesting why our concrete universe ended up the way it is, rather than otherwise. We consider such mathematical proofs a welcome hint at the possibility of early order making in our universe.

Continuous quantum fluctuations in so-called "empty space" are physical observations found at the core of quantum theory [22] [49]. From the point of view of abstractness, there is therefore no need for a distinct, separate and single act of creation: the sheer existence of abstractness and virtuality makes the eventual emergence of concrete universes inevitable. Concrete existence is the logical and inevitable consequence of virtual and abstract existence. This explanation nicely provides an "intangible" reason why, in the Cosmos, there is something instead of nothing.

Interestingly, this conclusion actually agrees with the religious contention that nothing can exist in the concrete universe to explain the fact that it contains something instead of nothing [55]: indeed, its ultimate source is in a different, abstract mode of existence.

Chapter 12: Universal Consciousness

The claim that we found a reason for the concrete universe to exist is an important conclusion with wide-ranging implications that we will inspect here in detail to make sure that it deserves our unwavering support.

To shortly summarise our deliberations so far: we assumed at the beginning of this book that "something" exists (p. 7), we reasoned that the concrete universe is finite (p. 17), that the quantum realm is timeless (p. 45), and that there exists a third, abstract mode of existence (p. 52). We also indicated how quantum particles and abstract knowledge are forms of energy, and we concluded that concrete existence is the logical and inevitable consequence of virtual and abstract existence.

A quick glance at abstractness reveals that it contains by definition an infinite number of possibilities; its general equations are completely abstract, they have parameters that accept an infinite number of values, and they cover all possible knowledge. At this point, it is important to note that "all possible knowledge" is the infinite knowledge of "everything that exists"; it includes, and indeed constitutes, self-knowledge, in agreement with the mathematical existence of a universal set (p. 57). Having said that, it is easy to realise that *"abstract, infinite, timeless and energetic knowledge of everything that exists"* constitutes the very definition of a *Universal Consciousness*.

This Universal Consciousness diffuses constantly into quantum virtuality in a manner that we interpret as the vacuum fluctuations of elementary particles. When the latter interact with each other, there occurs a truncation of their infinite values and they merge into the concrete world of space-time. In other words, the undeniable existence of abstractness implies the reality of a Universal Consciousness, its constant diffusion into quantum particles and the inevitable creation of a concrete universe.

It follows from this simple sequence that animal and human consciousness simply partakes in the primordial Universal Consciousness. We write the Universal Consciousness with an upper
case "C", but individual consciousness with a lower case to em-
phasise its partial, limited and finite nature. Our consciousness
is not purely abstract and eternal; it is virtually active and embedded in our brain and its network of neurons. Each of us can
only grasp part of the Universal Consciousness.

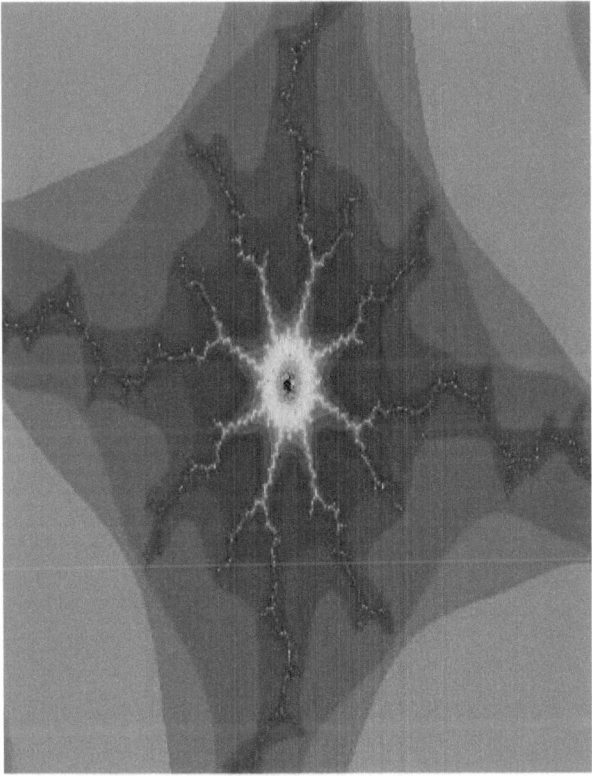

Figure 8. Fancy representation of a unit of consciousness in
touch with the rest of the universe. Called *"nousson"* from the
Greek *"nous"* for mind. Found hiding in the Mandelbrot set of
complex numbers at coordinates (-1.295952912,
0.441508883400000)

This view automatically solves the old mind-matter dichotomy problem. The French philosopher and mathematician René Descartes (1596-1650) promoted a dualism of two exclusive essences. Most modern philosophers now agree [56] that we are dealing instead with a dual-aspect monism, a term defined by the psychologist Max Velmans [57] to assert the unity of mind and matter as seen by science and common sense. We can now expand this notion further by saying that we are dealing instead with a triple-aspect monism incorporating the abstract, virtual and concrete modes of existence. The three modes are distinct but there are exchanges between them, and each mode cannot exist without the other two.

Figure 9. A wooden set of three where one cannot stand without the other two.

Defined as space-time, concrete matter is necessarily subjected to change (because time cannot exist without change) and therefore subjected to evolution (*i.e.* a series of changes). It is interesting to note that any concrete evolution necessarily involves an eventual return to the abstract mode of existence because of the interconnections between the three modes. That return occurs with the simple annihilation of matter-antimatter encounters, with the evaporation of black holes, and through the emergence of islands of consciousness after an obvious and ever more complex organisation of space-time. Dispersed matter thus has a natural tendency to re-assemble to its original abstract mode, like small mercury drops automatically rejoining into a single larger drop as soon as they touch. Accordingly - and despite frequent claims to the contrary - our concrete universe has shown a relentless tendency to increasing order and complexity, from plasma to particles, to atoms, molecules and organisms, leading to ever more complex information, consciousness and understanding.

This observation fully contradicts the pessimistic, yet popular interpretation of the second law of thermodynamics with entropy presented as a measure of universal and inescapable increasing disorder. The concept stems originally from what occurs in a steam engine producing work efficiently as long as it receives new energy. Left on its own in a closed system, any such engine inevitably (and literally) runs out of steam. Strictly speaking, entropy is a measure of the amount of energy present in a closed system and no longer available for doing mechanical work, and it inevitably increases in any closed system. This increase is often interpreted as *"increasing disorder"* but it does not need to be. On his educational web site, physics educator Frank L. Lambert asserts that this erroneous and persistent interpretation is *"as misleading as magic and as obsolete as 1898 fashions"*. To make it easier for physics students, he demonstrates that it is much easier to interpret entropy as a measure of spontaneous energy dispersal. Starting with the principles of the intrinsic motional energy of molecules and the probabilities assigned to their microstates, the notion of entropy explains how *"energy of all types changes from being localised to becoming dispersed or spread out, if it is not hindered from doing so"* [58], like radiating light, melting ice and the sound of music. Calling these phenomena "disorder" is not always appropriate, even if they have higher entropy and probability, like a disordered bedroom compared to a well ordered living room.

When applied indiscriminately to the universe, the "increasing disorder" interpretation has unnecessarily led to further confusion, pessimism and the loss of many bright and noble minds. The first thorn to be removed is that the second law of thermodynamics does not consider gravity because gravitational effects can be ignored in a steam engine. However, they definitely cannot be overlooked in the universe where gravity acts directly against entropy by bringing things together instead of dispersing them. Secondly, the earth and all planets in the universe are not closed systems since they constantly receive additional energy from their respective suns. The laws of thermodynamics certainly apply to the universe, but pockets of negative entropy are clearly allowed in billions of galaxies, at least for some considerable time. Finally, in the context of the three modes of existence, the concrete universe itself is not a closed system since it communicates with the other two modes of existence. In fact, the great idea of spontaneous dispersal also applies to the dispersal of abstractness into the virtual world of quantum physics.

To summarise this chapter so far, the notion of a Universal Consciousness makes sense if we

1) admit the existence of an abstract universal set and the possibility of infinite self-referral;

2) leave aside the old philosophical theory of dualism, and adopt the updated view of a triple-aspect monism;

3) embrace the notion that entropy is a measurable quantity of spontaneous energy dispersal, and abandon its mistaken and pessimistic interpretation as a measure of increasing universal disorder.

Matter may have started with simple subatomic particles and hydrogen, but these have been gathering ever since under the force of gravity or the structure of space-time, and have kept on aggregating into larger atoms, larger molecules, larger bodies that eventually organised themselves to the point where they became alive. Biological evolution then built upon this progress and reached levels of increasing complexity and manifest order.

Intelligent organisms eventually appeared, and then communities, societies and countries developed; we finally observe that apparent chaos is getting under control, not fast but steadily. Understanding and consciousness are obviously increasing on this planet, and probably elsewhere, at the leading edge of order making.   Consciousness thus appears to pervade the three modes of existence, and to extend its reach to any small piece of matter, inert or alive, like gravity does.

In 1988, Stephen Hawking made the claim that gravity was the major factor that would eventually determine the fate of our universe [43]. Gravity (which exerts its pull *in a snap*) certainly constitutes a major factor in the evolution of the universe, but more recent argumentation presented by science writer Charles Seife (in 2006) [39] and theoretical physicist Vlatko Vedral (in 2010) [59] convincingly assigns this determinant cosmic role to the production and processing of information. Scientifically, information was defined by the mathematician, engineer and cryptographer Claude Shannon (1916-2001) [60], with an equation identical to that of entropy and consequently with the same propensity for confusion. Its historical development is well presented by science author James Gleick in his book *"The Information"* [61]. Shannon was dealing with the type of information used in secret codes, telegraph messages and phone lines, and he defined it independently of any meaning it may convey. We can use his very effective approach to analyse the following coded string:

ABC BCD EBFGDHFG BIJC GEFH,

break its code and successfully render the hidden message as:

DER ERA SEOTAMOT EPIR TSOM.

Understanding the message's information was not part of his theory, and it constitutes a completely different kettle of fish. In our example, we extract the meaning when we read the answer from right to left and realise that *most ripe tomatoes are red*.

Obviously, understanding any information is more important than simply storing, processing or decoding it: chemistry, genetics and grammar books are wonderful, but understanding them is even more so. As a form of energy, consciousness depends on stored information and can locally appear much weaker than gravity. However, like gravity and information, it is cumulative and can reach very powerful levels on a large scale. It can be constructive or destructive, attractive or repulsive, positive or negative, but higher intelligence, argumentation and agreement can channel it into the same direction, so that for a sufficiently large amount of matter, the power of consciousness can eventually dominate all the other forces and determine the fate of the concrete universe.

The concrete universe did not spring out of nothingness (although it came out of nowhere) but emerged from the real, timeless energy of abstractness. It can be represented as space-time undergoing obligatory evolution, with its building blocks pulling themselves together and gradually waking up to consciousness. It presently looks like a gigantic jigsaw puzzle in the process of self -assembly. The more it expands, the higher its entropy, and the more conscious it becomes.

Chapter 13: Gödel's theorem

The growing importance of consciousness in the concrete universe, its pervasive role in the three modes of existence, as well as the fact that the world exhibits an internal order, all imply that the universe is rational, intelligible and understandable.

At first glance, this statement may seem to contradict logician Kurt Gödel's (1906-1978) incompleteness theorem [62][63], especially for those who interpret that theorem as a strict proof that the universe cannot be fully understood. In its original version, Gödel's incompleteness theorem is mathematically very intricate, but it can be translated into plain English without equations, with one of several versions proclaiming the following about mathematics:

In any formal system that contains at least arithmetic, there exists a proposition that we cannot prove (even when its truth or falsity is evident).

It follows from this statement that:

The consistency of such systems cannot be proved from within.

When first published in 1931, the incompleteness of formal systems was a severe blow to those mathematicians and philosophers who had until then assumed that everything could be logically explained. Under the leadership of the great mathematician David Hilbert (1862-1943), a plan had been drawn to prove all of mathematics based on a few axioms and assumptions, including that of internal consistency. By demonstrating that consistency could never be proved within any such system, Gödel put an end to their quest and drove several thinkers of that era to extrapolate their disappointment to other fields and conclude that the universe could not be fully understood. Understanding everything appeared impossible on logical grounds. This was the inexorable conclusion stemming from unsolvable paradoxes of self-referent arguments. It was based on the realisation that even if we expanded a formal system by developing it anew with more axioms and assumptions, its consistency still could not be proved from within the expanded system. We could expand to more levels *ad infinitum*, but the repeated attempts would simply lock us up into an endlessly iterating loop. This is very much in agreement with the finite nature of our concrete existence, and with the ZFC version of set theory. We must admit that some truths cannot be proved.

However, if we use the newer NFU version (Chapter 10, p. 57) with a universal set that contains itself, we realise that an endless expansion of formal systems is futile for the Universal Consciousness. Ultimately, we must admit that, by definition, Universal Consciousness certainly understands itself. Our finite nature may prevent us from knowing or proving everything, but everything should still make sense in the realm of infinite abstractness. Indeed, the lack of absolute certainty does not mean we do not hold the truth [64]. We may not be able to "prove" some propositions, but it does not matter when they are obvious or deliberately misleading. Their overwhelmingly high or low likelihood is all we need to respectively give them maximum support or no support at all. At the end of the day, we stand on firm ground as long as we support the hypotheses with the highest likelihood. After all, we do not need to count up to infinity to grasp the concept of infinity. We may never know for sure what our friends really think of us, yet this does not mean that the universe is incomprehensible. Despite his own proof, Gödel firmly believed that *"Die Welt ist vernünftig"*, the world is rational, intelligible and understandable! [63] This will also be our firm conviction.

Chapter 14: Emergence of concrete matter

One of the easiest ways to appreciate how concrete matter can emerge from virtual and abstract information is to have a closer look at the virtual reality produced by advanced computer games. Players wear special glasses, goggles or other headgear and are transported as by magic into a virtual reality where visual and tactile interactions can be as natural as in the concrete world. The best programs convincingly simulate imaginary worlds or real places such as an airline cockpit, the edge of a cliff or a meeting room. In some versions, you see yourself as an avatar interacting in a virtual world and the experience can be so intense that it can influence your behaviour when you return to your normal self. These computer techniques are now used to train soldiers in adverse environments or to organise virtual meetings with colleagues in other cities. These programs work because our brains interpret the "virtual reality" of computer games as they do the "real virtuality" that emerges from the abstract mode of existence.

Concrete and psychological reality is analogous to the "folders" and "documents" emerging from the simple bits encoded in personal computers. There are no carton folders and paper documents in personal computers; these folders and documents are an interpretation of stored information, just as a slab of concrete is, in the material world. At the atomic level, every solid object consists mainly of empty space, although our mind interprets that reality otherwise. Reality, whether virtual or concrete, is a mental construction. Call our world concrete reality, call it illusion or delusion, or call it a game, it remains the conscious interpretation of all the virtual and abstract information available to our senses and to our mind. The colours of the sunset, the softness of silk, the warmth of an embrace and the cruelty of war can all be expressed as mathematical statements in our heads or in computers, but they also constitute our reality. The universal set is necessarily an abstract concept, but it also contains everything that is virtual and concrete.

These ideas may be difficult to digest. However, the power of abstractness on the material world is not that mysterious. Consider that we experience it very naturally when we think of moving an arm and then move it. Something as simple as blinking on command demonstrates the power of thoughts over our concrete bodies. Recent research and experiments in neuropsychology have successfully demonstrated that monkeys and people fitted with appropriate electrodes can move a pointer on a special computer screen by simply thinking about it, and can also control specially designed equipment at will, by using the signals emitted in their brains while they formulate their thoughts. This is already helping people with spine injury regain some independence.

The mind interacts with the concrete world through the senses. What we see with our eyes, however, is far from being the picture we think we see. We now know that colour, movement, size, texture, etc. are perceived as separate inputs by the retina and relayed to our brains where they are interpreted and integrated into a comprehensive image. In vertebrates, there is a physiological blind spot where the optic nerve enters the retina, and yet, the brain recognises it and fills the portion of the field of vision that is not captured in each eye by extending the surrounding details and by using the information from the other eye to make a believable, complete image [65]. Most vertebrates are not even aware of the existence of this blind spot.

Amputees can often feel a phantom pain in the missing parts of their bodies, because of signals from the nerves that would normally extend all the way there. Countless experiments in neuropsychology have demonstrated that we can all be fooled by physically feeling, hearing and seeing things that only exist in our minds. When things go wrong with the brain, as in schizophrenia, affected patients can actually hear voices, see, feel, smell and taste things that do not exist at all. Some problematic dogs that keep on barking for no apparent reason may well be afflicted with equivalent, recurrent hallucinations. This is only possible because we already have a system in place to interpret virtuality into concreteness, and that system can be fooled.

Figure 10. Is this the graphical representation of a topological equation, a printed photograph, a two-dimensional PhotoShop drawing, or a three-dimensional paper origami?

The emergence of the virtual and concrete parts of the universe from abstractness occurs as it does in the examples above, as part of the interaction between the three modes of existence. Anyone willing to believe that "something exists" must logically admit that this emergence is natural, and that the concrete universe is a logical and indispensable extension of abstractness.

On a related note, it is almost comforting to see the widespread use of complex numbers in the equations of quantum physics. The concretely impossible square roots of negative numbers contained in the imaginary components of complex numbers no longer appear artificial or mysterious to those who accept the reality of virtuality and abstractness.

Amongst the infinite possibilities of abstractness, the simple existence of an optional clause allowing the truncation of infinite values is all we need to signal the automatic beginning of movement, time and evolution. The concrete universe is a mathematically truncated and finite universe emerging from the abstract and infinite Universal Consciousness.

Individual minds are partial echoes of that Universal Consciousness, and their attraction to it is intrinsically irresistible. Throughout history, innumerable Buddhist monks, Christian orthodox hermits, cloistered catholic nuns, shamans and mystics from all parts of the world have dedicated their entire lives to the contemplation of their versions of an abstract Universal Consciousness.

We conclude that, by completing the cycle linking the three modes of existence, the increased consciousness emerging from matter is slowly but surely putting the concrete universe in a position to control its own destiny.

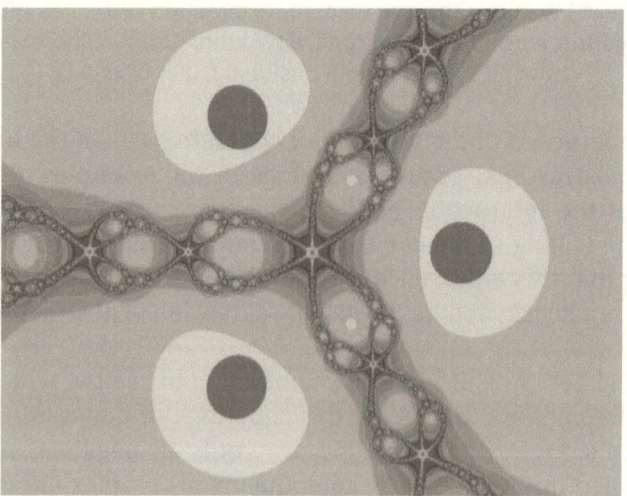

Figure 11. Computer generated graphic representation of Newton's method to find the cubic root of 1 in the complex plane. The central point is the origin of the real and imaginary axes. If we start the calculation at any point, the colour at that point represents the number of iterations necessary to find one of the three cubic roots.

Dialogue 6: Think a bit!

- Dog: After so many years of domestication, some people still doubt animals are conscious of what they are doing! It is discouraging!

- Monkey: I know, tell me about it!

- Lioness: We take care of our children better than many of them do, and they still refuse to see the obvious! We organise hunting parties and they seem to think we can do it without intelligence.

- Otter: They send their children to swimming classes, while we each teach ours on our own, without the need for specialists. I think they are dumb.

- Rock: Mmm!

- Dolphin, by the shore: All is not lost, my friends; several experts recently discovered that dolphins and whales have languages and even dialects. Some humans are beginning to think we may be conscious.

- Dog: I admit most dog owners speak to us, they know we understand and they consider us part of their families; I like that, but it's their scientific experts that get on my nerves.

- Monkey, rolling its eyes: Tell me about it!

- Lab mouse: They think we are not conscious because we can't speak; but no matter how much we squeak, they still don't understand.

- Cat: Will you stop all this noise? I'm having a nap here!

- Parrot: I can speak, but it doesn't help me much. My late cousin Alex [66] is one of my rare relatives who somewhat succeeded in making some people think we may be conscious.

- Rock: Mmm!

- Dog: As far as I am concerned, it's clearly a matter of degree: a squirrel and an engineer simply have different levels of consciousness, just as a weak battery and a power station have different levels of electrical energy.

- Angry squirrel: How insulting! I demand an apology!

Figure 12. Do people really mean we are not conscious?

Chapter 15: Randomness, selection and free will

The notion that consciousness emerging from matter may eventually control the destiny of our concrete universe implies free will, one of the most contentious and controversial concepts in the history of religion, philosophy, psychology and politics for over two millennia. We will certainly not settle the discussion here, but we will point out that the early proponents of strict determinism who claimed there was no free will were not aware of the probabilistic properties of quantum mechanics, and based their conclusions on an incomplete understanding of classical physics. However, if free will is incompatible with strict physical determinism, it may also appear to be at odds with the randomness inherent in the virtual mode of existence.

Randomness may suggest an anarchic process out of control but, on the contrary, it stands on solid ground because the laws of probability are very precise and, paradoxically, do not leave anything to chance. Indeed, it is always possible to generate certainty out of randomness. Telecommunication, the internet and a myriad of electronic devices are as many proofs demonstrating the accuracy of random quantum effects. The same is true in the classical sphere: for example, flipping a coin gives heads 50% of the time. Flip as you may, your results generally approach closer to 50%.

It is worth noting here that randomness, like space, time and speed, is a relative concept. If you pick up numbers at random, 19 is a random number, but it is no longer random when it is your age or your street address. The reverse is equally true: a non-random sequence can be made into a random one. For instance, any particular sequence of the infinite expansion of π (pi, the ratio of a circle's circumference divided by its diameter) can be used as a random sequence in a guessing game, even if the sequence of digits in the expansion is fixed. We could start the game at the 29th decimal, right after 3.1415926535897932

384626433832, and most players would successfully guess the next digit only once out of ten, in full agreement with the laws of probability. A sequence of digits or events that would be absolutely random under all circumstances simply does not exist. The simple truth is that a sequence is random if one cannot determine it in advance, and it is not random if one can.

Despite some people still claiming that the world is completely deterministic, we see randomness everywhere. Biological evolution is based on random mutations spreading through populations following natural selection of the fittest, while some genes disappear or become predominant following random drift. An example of random drift is when a sudden flash flood wipes out an entire animal population with the sole exception of a mating couple who happened to be on high grounds. Their genes will be inherited by all subsequent generations not because they were the fittest but because they were the luckiest. This example shows that selection is far from being unidirectional or strictly deterministic; like genes that mutate at random, selection is also subject to random changes. Selection in a desert will favour those who can retain more water in their bodies, but when the climate changes and gets very wet, selection will move in the opposite direction, as an essential component of change.

Let us also consider the importance of randomness and selection in other biological processes. During cellular division, random exchanges between pairs of chromosomes produce new gene combinations, some of which are selected during the spermatozoa's race to insemination. During embryogenesis, neurons develop according to the plan coded in the genome, but their arrangements and interactions are established at random and it is through a selection mechanism of useful dendrite connections versus useless ones that we can eventually think and remember. Environmental factors such as maternal nutrition and the chance exposure to teratogens also play a role during fetal development. The fight against foreign pathogens is assured by random splicing of immunoglobulin genes followed by selection of those cells that produce the best antibodies against particular invading antigens. Blood vessels around the intestines also develop randomly but their final network reflects the selection of those associated with the minimum of blood flow resistance. Randomness and selection are essential everywhere in biology!

Figure 13. Randomness and selection.

In real life, we seem to thrive on randomness and uncertainty. We spend billions of dollars annually on lottery tickets, in bingo halls and casinos. We all play cards, snakes and ladders or other games of chance. More professionally, consider stock market surprises, risky business deals and crop losses due to sudden storms or earthquakes. It is again obvious that randomness and selection play fundamental roles in the universe.

In practice, strict determinism and randomness intermingle profoundly and effectively at all levels. The mixture is universal and very much compatible with free will as we demonstrate below with the three following examples:

1) the law of large numbers in statistics,
2) the Hardy-Weinberg equilibrium in genetics, and
3) chaos theory everywhere.

1) The law of large numbers

The law of large numbers is a mathematical law, not an un-proved theory or an uncertain hypothesis. It states that large random samples of any population will converge to the aver-age of that population regardless of the variables' underlying distribution. The distribution of a measurement in an entire population may look like a bell-shaped curve on a graph, with most measurements falling in the middle (where the average is), less measurements on either side of the average, and very few at the curve's extremities. Other types of measurements will follow distributions with averages closer to one end of the range of measurements or with two or more peaks on the graph. But for every distribution of any shape, there is a popu-lation average; and what the law says is that calculating the av-erage of a large random sample (even if only from a small per-centage of the population) will give us a very good estimation of the entire population's average. If we take several random samples, their averages will also converge to the entire popula-tion's average. The important word here is "random". If we try to estimate the average salary of people in North America by looking at the paychecks of the most famous hockey players or Hollywood actors, we will not obtain a good estimate be-cause the sample is not random. Making sure it is random re-quires a significant effort but allows us to find the population average without having to verify the salary of every single per-son on the continent. In this case, randomness is far from lead-ing to anarchy or unpredictability: coupled with a very precise statistical law, it leads instead to accurate knowledge.

2) Hardy-Weinberg equilibrium

Random mating occurs when animals or people reproduce without any consideration to a particular characteristic. For example, people do not generally get married on the basis of their blood groups, which is why we can say that mating is random with respect to blood groups. On the contrary, mating choice is not as random when we consider external physical features. All inherited characteristics are determined by genes, and genes exist in various forms called alleles. When all members of a population can reproduce and survive, it is remarkable that the proportions of alleles in the population remain constant from generation to generation, whether mating is completely random or strictly assortative. When mating is random, the genotypic proportions (allele combinations in individuals) also remain constant in the population. This principle is known in genetics as the Hardy-Weinberg equilibrium [67] [68]. In this case, randomness combines with a strict law to allow freedom of choice within a constant order.

3) Chaos theory

Chaos theory [69] started officially in 1963 when meteorologist Edward Lorenz (1917-2008) plotted the strange results of three simple equations. After him, the mathematician Benoît Mandelbrot (1924-2010) and many others working on computer models discovered the same features in all areas of science. What they found is that when the parameters of non-linear equations increase beyond certain critical values, the answers to the equations no longer converge towards single values but start instead oscillating between two values, then four, and eventually jump all over the graph in an orderly, yet unpredictable fashion. The name "chaos" here is a misnomer. It turns out the convolutions of smoke rising in the air, the complexity of cloud formation, wind turbulence or spurts of growth can all be explained by the recurrent application of precise equations producing what proves to be only an "appearance of chaos": a blend of precision and randomness, a puzzling mixture of order and disorder. Applying a non-linear function to any system, and re-applying it to the result obtained, again and again, creates a pattern of self-similarity at all scales, *"like stretches of the map of Britain seen from different altitudes"* (an example often quoted in this field). Self-referral and self-similarity are seen in the development of ever smaller branches in trees and ferns, in neural networks, blood vessel distribution, the formation of mountain ranges, snowflakes, waves, galaxies, it is found in physics, chemistry, physiology, genetics, cytology, industry, economics, music and everywhere else in nature and the universe. Plotting these mathematical operations in two or three dimensions gives rise to wonderful pictures known as fractals [69] [70] [71] such as the computer generated illustrations found in this book. Chaos gives to the Universe its appearance of organised randomness as reflected in living organisms, in scenic beauty and musical symphonies. This harmonious and beautiful blend of precision and randomness makes the world the exciting (non-boring) place that it is. We now know that most of Nature is regulated by these dynamic systems, which are locally unpredictable despite being globally deterministic. In this case, again, the pervading presence of randomness and uncertainty throughout the universe allows us to reject an exclusively deterministic world where every single event is preordained from the start.

While Darwin's theory of evolution emphasises the chance selection of random mutations, we deliberately select engineered mutations in the food industry and husbandry, and we democratically steer the course of social evolution. Obviously, free will is not absolute, as we must exercise it within the limits of our nature. We cannot fly like birds or swim like fish, and we are conditioned by our genes, our prenatal environment, and by parental teaching, but despite reasonable restrictions, the reality of at least some free will is undeniable. Claiming that we have no free will based on the accurate observation that conscious decisions are taken subconsciously hundreds of microseconds before we become conscious of them [72], is like our claiming that you did not clap your hands because you were far away and we only heard you clap a second later. Indeed, both our subconscious and our consciousness belong to us, and they work as distinct but complementary modules, like the separate optical inputs that get integrated within our brains to form clear images (as already mentioned on p. 74). The mind is embedded in our physical brains but it must also have a virtual component. It is granted that all our decisions are influenced, and sometimes strongly conditioned, by our upbringing, but concluding that they merely appear *"as sprung from the void"* [73] is a bit of an exaggeration, unless we equate void and virtuality. In fact, we can alter our social conditioning through politics, education and activism to improve society's free will. Some societies are permissive and others autocratic, but in the last analysis they are what we make them. In reality, the more conscious we become, the more free will we acquire.

We conclude that there are restrictive and chaotic influences to our behaviour, but with plenty of opportunity for choice and liberty. We can confidently refuse and avoid fatalism and submission. We can stand up and give meaning to our lives. The concrete universe can wake up and select its own destiny! Given any degree of free will, the emerging consciousness of the concrete world will choose naturally and deliberately to develop itself further. Current attempts include instruction, education, libraries, scientific research, the internet and the construction of quantum computers, the eventual consciousness of which is as inevitable as our own.

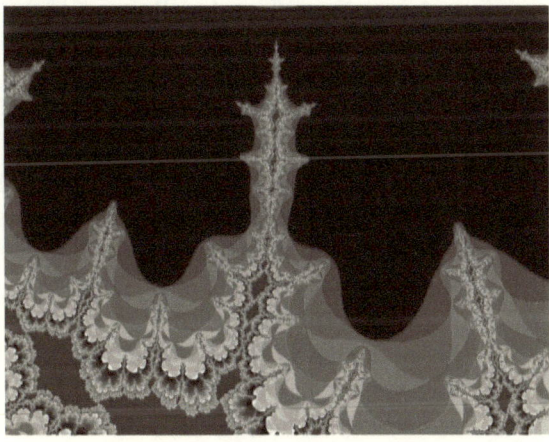

Figure 14. Part of the universe waking up and standing up. The complex function used to calculate and plot this fractal is related to that of the Mandelbrot set: $z = 2z^2+c$, with upper left corner coordinates at (-1.2725937805175670, 0.0053160705566407).

Chapter 16: The essence of our existence

The previous chapters and dialogues provide solid scientific support to the theory that the deep basis of our existence is its underlying Consciousness. This is hardly a new idea: every major religion and every major school of thought in history has taught and teaches the unity and supremacy of an abstract principle. Across cultures, Consciousness is perceived imaginatively as power or as the source of existence, sometimes in esoteric mysticism, but also philosophically and scientifically, with elegance and convincing argumentation [74] [75]. In keeping with this tendency, we needed very little mental effort to accept Consciousness as a form of physical energy (see Ch. 8, page 49) on equal footing with electricity, magnetism and gravity. As we should have expected from the start, our physical theory of everything must now blend with the theory of the mind in psychology.

The Copenhagen interpretation of quantum physics deliberately avoids its ontological implications. In contrast, David Bohm's (1917-1992) version embraces them while still producing the same mathematical answers. This led Bohm and Hiley [76] to claim that *"everything is strongly and nonlocally connected"* and that *"quantum theory contains a 'classical world' within it"*. With this vision, they concluded that the physical and intangible aspects of reality are two inherently related aspects of a greater order and they joined the universal movement of philosophy, arts and religion that sees an intangible interconnection between all things, from Parmenides' assertion that everything is One, through the poet William Blake (1757-1827) who invites us

> *"To see the world in a grain of sand,*
> *And a heaven in a wild flower,*
> *Hold infinity in the palm of your hand,*
> *And eternity in an hour"* [77],

to the Dalai Lama who sees the universe in a single atom [78]. The conclusion is everywhere the same: *"the Universe is whole, ordered and conscious"* [74].

The essence of our existence thus implies a sense of comfort with order and unity, and a natural tendency to share and co-operate. This conclusion is at odds with the popular concept of the selfish gene [79] and its extension into the belief that selfishness rules the universe. In fact, for optimists, nearly all arguments made in support of the selfish gene can be reversed and made into arguments in favour of co-operative genes. A gene left alone on a table is a dead gene. To survive, it must be part of a complex machinery that includes many other genes collaborating to keep the cell or the organism alive and to reproduce it. It is true that parasitic genes can hitchhike a free ride at the expense of other genes [80], but when one stops on the road to give a lift to a hitchhiker, one demonstrates generosity and co-operation, not selfishness. A human hitchhiker often turns out to provide pleasant company and may unexpectedly be very useful in difficult circumstances (although there are unpleasant exceptions). Similarly, in biology, the genome that accepts hitchhiker genes can be viewed as co-operative. In fact, the addition and further mutations of originally extra "junk" genes often fuel the evolution and survival of species (and the occasional disease).

The genomic set-up itself provides further evidence of co-operation: the presence of the entire genome in every nucleus of the organism assures full information at the lowest level, and is a reflection of the presence of infinite probabilities in every particle/equation. There is also further co-operation between not less than three genomes in every cell: the chromosomal haploid copies of paternal and maternal origin, and the maternal mitochondrial genome involved in energy management. Symbiosis with our indispensable gut bacteria, and our societal organisation (with each individual aware of the societal rules) are further evidence of our deeply co-operative nature.

With such a hospitable background, we should expect all major forms of pleasure to involve sharing and co-operation. On logical grounds, if our nature is essentially co-operative, a personal participation with the rest of the universe should naturally convey deep feelings of happiness, fulfilment and satisfaction. If we find that this is the case, we will gain more support for the views developed above. Happy feelings can be represented by a child laughing or an adult smiling contentedly in the sun, but let us concentrate on more intense examples. Let us identify those memorable situations where people report or display a climax of intense happiness accompanied by distinctive feelings of magical enchantment and sublime fulfilment, those unforgettable moments summarised by snapshots of people with eyes closed, holding their breath and ready to faint, or with eyes wide open, fists clenched, shouting victory to the world. If we do, we end up with a list as follows (you can prepare your own list if you want):

> Amorous passion
> Religious ecstasy
> Musical rapture
> Parental devotion
> Sexual orgasm
> Intellectual elation
> Artistic fervour
> Tribal exultation
> Sports euphoria
> Political jubilation
> Absolute certainty, and
> Mission accomplished.

When we inspect the list attentively, we realise that every single item implies and requires sharing and passion. In their most successful forms, all these examples of extreme happiness require a partner, a community or the rest of the world to share with. If we try to find happiness without sharing, without love and compassion, we immediately fall into selfishness, exploitation, hypocrisy, dishonesty, mediocrity, cheap art, rape and war.

The pursuit of real happiness cannot be a selfish goal; we can only find it if we are in harmony with the rest of the universe. Once we understand this at the personal level, our next step is to teach these principles to our children and to those around us. By leaping this way from local to regional to national and international levels, we can help the universe get closer to its final destiny, at least in this part of our galaxy. As far as the universe is concerned, this is certainly one more step in the right direction.

We have come a long way from studying the physics of an iron rod and of quantum mechanics, to discussing happiness and contentment. Perhaps surprisingly, all those subjects turn out to be interconnected. When joined together, the principles we have been lining up are now looking more and more like a rising theory of everything.

Chapter 17: Cultural leaps

In the periodic table of elements, the discrete levels of increasing energy that surround the atomic nucleus are gradually filled with electrons until higher levels can be started. Going up to higher levels occurs in an orderly fashion, after the completion of all preceding levels. Moving an atom from one level to the next is a discontinuous and abrupt motion that physicists call a "quantum leap" (or quantum jump); after a leap, electrons do not stay long at high energy levels if all the lower levels are not first filled with other electrons.

This very simple and primordial system of layered organisation is found everywhere else in the Cosmos, in everything that grows and develops. The lower levels are established first, and the higher levels follow. Unfortunately, as in atoms, we cannot prematurely leap to higher levels: we must grow in stages.

Examples of layered growth include bypass roads built around pre-existing cities to improve traffic. In sports, international associations rise only after local clubs and then regional and national leagues are formed. Hospitals can only function well when all their constituent departments do, including the emergency, surgery, obstetrics, laboratories, finance, housekeeping and building maintenance departments. Cities can only function when they possess the necessary infrastructure including police and fire departments, hospital, emergency services, transportation services, post office, library, stores, recreation facilities and a well-organised City Hall. Once we have cities (or in history, city-states), we can unite them and build countries. The United Nations only started after most countries first recognised each other as independent; and it will only function properly when superpowers are forced to relinquish their veto rights and stop acting at two levels simultaneously. In this light, U.S. President Bush's insistence on suddenly exporting and imposing democracy to Iraq and Afghanistan without first establishing the social prerequisites and infrastructure appears to have been sadly misguided.

Life itself works in the same layered way: first molecules that can reproduce themselves, then cells to protect these molecules; and then cellular co-operation to form multi-cellular organisms. Living organisms cannot exist without the prerequisite smaller cells and a lot of basic biochemistry. It takes time for them to grow up and the same is true of social organisations. Even when set up and managed by very well informed and educated people, organisations need to grow and go through the equivalent of childhood and adolescence before being successful. We are all too painfully aware of associations, public services, cities, ministries or entire countries that have not yet reached adulthood and make life difficult for their upset citizens.

Attempting to work at higher levels before securing lower levels does not make sense and is a recipe for failure [81] [82]. This difficulty is even more evident when one considers the gradual evolution of ideas: those ahead of their time are not only ignored but they invariably meet with social resistance. Just as electrons must first fill a lower level before starting on a higher one, so it seems with ideas, whether scientific, philosophical or religious.

Consider for instance the crisis that faced Pythagoras' (ca. 570-480 BC) school of thought in ancient Greece when it discovered the irrationality of $\sqrt{2}$ (the square root of 2). The finding shattered the belief that everything could be explained in terms of whole numbers and their ratios. We are told that the finding was first kept secret and forcibly suppressed.

Aristarchus had already concluded in antiquity that the earth had to circle around the sun, rather than the opposite, but the layer of knowledge achieved at that time was not full and his ideas had to wait 1,700 years to flourish.

Figure 15. Aristarchus of Samos
(310-230 BC)

In the 14th century, John Wycliffe criticised abuses and false teachings in the Catholic Church. While alive, he was denounced by Pope Gregory XI; after his death in 1384, Pope Martin V ordered his remains to be exhumed, burned and the ashes thrown in a river. Reformation had to wait for nearly 200 more years.

Remember what the Catholic Church did when Copernicus (1473-1543) and Galileo (1564-1642) proposed (after Aristarchus, Indian and Arab astronomers) that the earth was turning around the sun. Once more, the official reaction was negative and harmful, as it involved stubbornness and repression. Society was not ready.

Then we have Newton's (1643-1727) laws that revolutionised physics but were stubbornly resisted in France until alternative explanations were proved wrong. In all cases, the problem is that new theories disturb established views and originally seem to bring chaos instead of order.

Figure 16. Mary Wollstonecraft (1759-1797)

During the French revolution, Mary Wollstonecraft, a writer and early feminist (whose daughter Mary Wollstonecraft Shelley is the author of *Frankenstein*), dared argue in England that slavery was immoral, that women should have the right to vote and that men and women should be entitled to the same education for the benefit of all humanity. Her views and lifestyle shocked society. It took 200 more years for the feminist movement to flourish in the West, but despite this evolution, most women in the world today are still waiting for full emancipation.

The theory of biological evolution is following the same path. Charles Darwin (1809-1882) himself kept it private for decades, as he did not think time was ripe for publication. More than a hundred years later, a few American states still make the teaching of evolution illegal in school.

Then we have Mendel's (1822-1884) laws of inheritance, which were not fully appreciated until after his death, when the scientific community finally got ready to digest them.

The theory of quantum physics is another big leap, resisted even by Einstein. Although all physicists now accept its rules, most have not yet come to terms with its profound philosophical, religious and psychological implications, because it touches on all aspects of life, society and ethics. It always takes so long to change outdated ideas that there is little hope of seeing the worldwide celebration of a successful Grand Theory of Everything in our lifetime, as already discussed at the beginning of this book.

Developing a scientifically sound theory of everything no longer requires new data. We are presenting our own version in this book, and we strongly feel that many people, worldwide, are ready to consider that time has come to leap to the next layers of understanding. Meanwhile, the majority will unquestionably take much, much longer to adapt and to adopt more modern views. It will take several generations for humanity to reach adulthood, but progress is unrelenting, and we should at least start spreading the good news!

Chapter 18: The Grand Awakening

In the previous chapters, we reflected on the fact that the concrete part of the universe is evolving towards more consciousness and that it should eventually manage to understand enough to take its future in its own hands and do what it is meant to do. Like a child who does not yet know what it is going to do when it grows up, the concrete universe is at an early stage of its evolution: it is literally waking up to consciousness (at least, this is what it looks like to us from our point of view in this part of the universe). We can compare this awakening to a glass of water put to boil on a hot plate. After a while, we see small bubbles appearing on the sides of the glass, apparently out of nowhere. Some bubbles eventually detach and rise to the surface. Soon, bigger bubbles form, group together and cause more movement. Finally, the water reaches the boiling point and becomes very turbulent. As Matter gets organised and wakes up, animals are its first small bubbles, we are the equivalent of the somewhat bigger bubbles while higher intelligence is getting ready to reach the boiling point.

We do not participate in that Grand Awakening as foreign observers or embedded reporters or as independent players or even as co-participants, but as Matter itself. We are made of matter and are not even a separate, distinct part of it since the atoms and molecules that form our bodies and brains at any instant are continuously exchanged for new ones, in and out, as we breathe, grow, eat and burn energy to live, to think and to understand. Recycling and replacing parts of ourselves at every moment, we always remain by nature an integral part of that evolving universe.

Intuitively or scientifically, we all agree on our participatory nature, as expressed by the visionary Carl Sagan (1934-1996) in one of the catching pronouncements in his book *"Cosmos"* [83]:

> *"The nitrogen in our DNA, the calcium in*
> *our teeth, the iron in our blood, the carbon*
> *in our apple pies were made in the interiors*
> *of collapsing stars. We are made of*
> *starstuff."*

We find the same intuitive feelings in the bereavement poem first drafted in 1932 by Mary Elizabeth Frye (1905-2004):

Do not stand at my grave and weep

> *Do not stand at my grave and weep;*
> *I am not there; I do not sleep.*
> *I am a thousand winds that blow,*
> *I am the diamond glints on snow,*
> *I am the sunlight on ripened grain,*
> *I am the gentle autumn rain.*
> *When you awaken in the morning's hush*
> *I am the swift uplifting rush*
> *Of quiet birds in circled flight.*
> *I am the soft stars that shine at night.*
> *Do not stand at my grave and cry,*
> *I am not there; I did not die.*

From this participatory perspective, it appears that ever since its first appearance more than a hundred thousand years ago, humanity has played its part in a grand scheme without knowing it. But now, at last, we found out what we are doing here; we finally reached the point where we can consciously march forth with the rest of the concrete and awakening universe to turn the odds in our favour and increase the universal development of consciousness and understanding.

Note very carefully that all the above naturally implies the eventual development of an intelligence much greater than ours. Future intelligent beings will be so much ahead of us that they will not call themselves humans, more than we call ourselves monkeys. Realistically, it clearly means that we are not the end of the world, but to be frank, it is hard for many of us to swallow and digest so much humble pie.

In keeping with the above, we do not agree with the sophists (Protagoras, *ca.* 490-420 BC) that *"man is the measure of all things"*, with the white queen that *"the whole world is a dumb chessboard"*, or with the existentialists (Sartre, 1905-1980, de Beauvoir, 1908-1986) that the world is absurd and meaningless. We clearly are a simple link in a long chain that is far from completed.

The next dialogue is an illustration of the intellectual danger that lurks in the background when people think they are the centre and apotheosis of the world.

Dialogue 7: Evolution

Her Majesty the Queen was resting in her comfortable palace, as usual. All was well, the kingdom was at peace, the food aplenty, and the ants happy. Laying all those eggs was quite a chore, but it had been a long time since her last brood. The ant workers really took good care of everything, and she was now completely free to do as she pleased, as she ordered.

- Guard! She called.

- At your order, Your Majesty! Replied the royal guard immediately, coming out of the shadow.

Betty was a new royal guard and had recently made her way to this enviable position. She was good looking and could be ferocious when needed. Today was her first shift in full service and she was determined to make a good impression. Otherwise, she may be demoted back to the kitchen! But the Queen was in a good mood, as usual. In fact, she had a wonderful reputation and was known to like the arts and enjoy modern folk songs, of the kind you sing while carrying food parcels to the nest over long distances. She also liked science and philosophy and had taken a keen interest in listening to her two favourite specialists debating about evolution and the meaning of it all.

- Guard, she asked, please inform Jean and Julia that I wish them to come and dine with me this evening at six.

- Done, Your Majesty!

Betty saluted with her antennas and left right away, wondering who on earth were Jean and Julia. So she headed straight to the information desk where Old Mary was yawning.

- How can I help you, young girl?

- I have to find two ants named Jean and Julia for the Queen. Would you know where I could find them?

Old Mary rolled her eyes and sighed.

- In the B wing, of course!

- In the B wing?! But that's the male wing! I said "Jean and Julia", Mary!

- Oh, Betty, don't tell me you don't even know who Jean and Julia are! Mary replied and sighed again. She then took a deep breath and slowly explained:

- Jean and Julia are males. Male black ants, not red like us. They were enslaved by mistake with a larger group of black worker ants when they were still larvae. They are totally useless, like all other males in our colony, but when the error was discovered, the Queen thought they were cute and decided to keep them for educational purposes. Since they don't do anything all day, they had plenty of time to educate themselves. I really don't see why we don't chop their heads off, but the Queen is the Queen.

Betty's eyes and mandibles were now wide open.

- Come on, Betty, go! The Queen won't wait!

Still stunned, Betty half turned away on three legs and asked:

- But why do they have girls' names?

Mary paused and thought.

- I am not sure, she said. People generally assume the nursery department named them when all the abducted larvae were thought to be females and that the names stuck to them. But none of those involved at that time has ever confirmed that with me. I suspect their own people baptised

them and that in their native language these are boys'
names. They belong to another species, remember?

Figure 17. Fire ants.

As she made her way towards the B wing, Betty wondered what
other surprises were awaiting her there. That new job was so different! When she arrived, however, there was no need to ask any-

one, as there were only two black males in the drones' chamber. All the other males were red, as they should be, and they looked silly, Betty thought. After the introductions, Julia formally announced they would both be pleased to attend dinner at the palace at six. Jean was the dreamer type and simply smiled warmly. At six, the two showed up at the palace door and were ushered in. They were wearing their best perfume, smartly covering the smell of their own stinky pheromones. Foreign ants smelled so bad it was always easy to detect and attack them in a crowd, even when their species were closely related and looked almost the same. Purity of the race is a sacred principle among ants!

- How nice of you to have come! Rejoiced the Queen.

- Your company is always a pleasure, Your Majesty, replied Jean honestly.

- Let us sit at the table, she continued. While the meal is being served, please tell me more about the theory of evolution you mentioned last time. It sounded quite interesting but hard to understand. Start again at the beginning, if you don't mind.

- Jean no longer calls it a theory but a physical law that rules the entire living world, noted Julia judiciously.

- Oh yes, Jean exclaimed, I find it so fascinating I have been dreaming about it all week.

- Go on then, tell me! Said the Queen, all antennas.

- It is one of your cousins' ideas. She studied inherited variations among and between several hundred species of ants and concluded that we are all descendants of a prehistoric type of ants that lived several million years ago.

- Nonsense! Objected the Queen.

- That's what we first thought too, obliged Julia, but we eventually changed our minds in the face of her overwhelming arguments.

- What arguments?

- Small mutations in a Queen's genetic makeup are favoured by chance in certain environments. The new characteristics they determine are selected in her offspring and passed on to the next generations until they distinguish the entire population. Since all of a nest's population comes from the Queen, new genomes can get established in very few generations and quickly lead to the emergence of new species unable or unwilling to interbreed with others. When the mutations affect our ability to produce or smell pheromones, the resulting differences are stronger than any brick wall or ocean between two species.

- Amazing!

- This is why we, ants, have been so successful at dominating all the animal kingdom. With thousands of species covering the entire planet in all possible environments, temperatures, altitudes and latitudes, we are absolutely invincible as a group. We can lose hundreds of species, and yet some species will survive any cataclysm and others will quickly evolve under any new condition. Add to this your enlightened administration and you can see why you find yourself at the top of the evolution chain!

- But we are such small insects!

- Ha! That's what the dinosaurs thought too! Have you seen any of those big bragging brats around lately? Scorned Julia. Each ant colony is small, but our total biomass makes up 10%-20% of all animals presently alive on earth [84] [85].

- What about the human animals? Their technology looks rather advanced, don't you think?

-   With all due respect, Your Majesty, humans are doomed. Doomed for a multitude of reasons. The most obvious is easy to understand: we have tens of thousands of species, while they only have one! Any slightly successful type of animal would have already succeeded at evolving into at least a few dozen species.

-   There is only one species of humans?!

-   Yes, all humans smell the same, they all taste the same and they can all interbreed! That's how a species is defined: a group of individuals that can successfully interbreed. In terms of evolution, they are deep at the very end of a dead end, Your Majesty! The next mutant virus, the next ice age or severe global warming can wipe them all out in a few hundred years. Their kind is hopeless!

-   Really?

-   Obviously! If that was not enough, they also keep on exterminating each other. They probably got rid of the Neanderthal race on their own, a significant percentage of their yearly national budgets goes into warfare and they kill each other in the hundreds of thousands every year, with only a tiny proportion of the losers kept as slaves. They are dumb: despite this grim outlook, their superpower states all promote and still use the death penalty. Their latest discovery is the genocide!

-   Well, what about their so-called intelligence?

-   The more educated they are, the less they reproduce. Worldwide, the less advanced have more children. They are evolving backwards!

-   Incredible! How did they avoid extinction so far?

-   Sheer luck, Your Majesty, sheer luck!

- But it won't last long, added Julia. Were they to amend their ways through reasoning, they would still be stuck with their dumb mode of reproduction. You may not believe it but every female can reproduce with any male! Think of all the waste of energy spent by each human couple in their efforts to bring up a tiny brood of only a few babies in a lifetime, and compare it to our efficient method of producing thousands of new ants each year, cared by the thousand progeny from last year. Their method is sheer nonsense. The last blow to their optimism, if they have any left, is that their Y chromosome is now known to undergo inevitable length reduction at both ends, known as telomeres, so that in several thousand years, they will no longer be able to reproduce anyway [86]. Believe us, Your Majesty, ants are the most successful creatures and the most advanced!

Betty had been standing straight and motionless during the entire meal. She could not believe it. Despite being males, these two weirdoes made a lot of sense and were most likely right.

Her Majesty was pleased.

Chapter 19: Science and Religion

We do not agree with everything the ants said in the previous dialogue, although their criticism is well taken. There is no valid reason for humans or ants to think they are the ultimate jewel of creation. At the very least, if we acknowledge the reality of evolution and the high probability of a relatively long future, we must admit that in many, many million years, evolution will have stayed its course and the intelligent folk of the time will see us as we now see prehistoric ants. Or worse! In other words and as previously mentioned, we are not the end of the world!

We demonstrated in the previous chapter (p. 95) how the concrete universe is slowly awakening to consciousness and gradually organising itself. The eventual goal of this evolution, if any, is not immediately obvious. Humanity has been trying to figure out that goal and the source of its existence for a long time. In their search for truth and understanding during eons of observation and reflection, all the world's religions, philosophies and cultures have described the ultimate explanation in their own words as the Hindu Nirvana, the ancient Greek Elysian Fields, the Chinese Tao, the Judeo-Christian Heavens, the Islamic Jannah, the Buddhists' Enlightenment, the Viking's Valhalla, Teilhard de Chardin's Point Omega [87], Ray Kurzweil's Final Singularity [32] and many, many others. From various angles, they all preach – or preached – their partial understanding of the world, whether prophets, clerics, philosophers, wise elders or scientists.

In his book entitled *"The blind spot"*, mathematician and statistician William Byers writes [64] that science and religion are perfectly compatible *"for ultimately they are the same thing"*. Both search for the truth and try to explain the source of our existence and give it meaning. Being better informed than in the past, science is often called the religion of our time, and it gradually provides more and more explanations for phenomena that used to be superstitions or religious mysteries. It is pertinent to remember that religion was a normal step in the evolution of every civilisation, and that organised religion still plays an essential role in the teaching and development of science in many societies. At the dawn of civilisation, religions provided answers to most questions and promoted social cohesion. As human groups increased in size, organised religion further fostered co-operation and civil obedience, thereby assuring competitive advantage. The records of human civilisation show that religion and science were complementary rather than opposite throughout history, with organised religion being a driving force for spiritual and scientific investigations, while science gradually grew in importance as time went by. No one denies that there have been clashes between the two (as well as within religion and within science), but it is unfortunate that in Western society today (in contrast to what we see in Eastern cultures), science and religion are increasingly and unnecessarily perceived as brother enemies. There is no denying that some sects, cults and faiths can preach utter nonsense [88], but some congregations are extremely reasonable and recognise the authority of science [55] [78]. Both science and religion are human activities, and it is unavoidable that representatives of both camps occasionally adopt very rigid stances incompatible with their respective basic principles.

Abruptly getting rid of religion is not an acceptable alternative to all those who live by their faith and require the loving support of their religious communities to give meaning to their lives. Telling these people they are "deluded" [88] cannot convince them they are making a serious mistake (even when they are), especially when the tone used is perceived as arrogant or hostile and when atheist alternatives have no welcoming and active communities to turn to. Belief in a personal god may be a delusion, it may have fuelled wars and extreme chauvinism, but it also continues to inspire the highest forms of brotherly love, honesty, kindness and benevolence and will not be abandoned without an equivalent or a more attractive and positive alternative. Until we have a widely recognised and universally accepted theory of everything that deserves more support than any other theory, there will be plenty of room for religions and alternative scientific theories.

It is important to realise that regardless of which cosmology we embrace, we all believe in something we cannot prove conclusively, be it a creator god or goddess, a creator couple, a generating golden egg or feathered serpent, a primal coconut root, a constant creation to justify a steady state universe, an inflation theory to justify the Big Bang, or the triple-aspect monism of reality. When we take all this into account, we realise that most people will stick to their blind faith or will simply keep an open mind for as long as they are personally satisfied with their own explanation of why, in the concrete universe, there is something instead of nothing.

Whenever a religion preaches nonsense, it is not a god that is to blame, but that god's ignorant, mistaken or troubled representatives. Eventually, it all comes down to applying the method of support to the competing explanations of science and religious organisations by comparing their respective likelihoods and deciding which deserves higher support [89] [90]. Amongst them, the ultimate explanation presented in this book is a contender that deserves thoughtful consideration: concrete existence is the logical and inevitable consequence of abstract and virtual existence.

Predictably, many will be tempted to say that the Universal Consciousness discussed in the previous chapters is nothing more than what they already believe in, but this may necessarily require serious adjustments to the details of their theologies or philosophies. In particular, they would have to distance themselves from mistaken religious leaders who call for armed "holy wars" or who pretend that their omnipotent gods have personally given them exclusive rights to a particular land or insist that they dress in a particular way or eat particular foods, cooked in particular ways. In fact, no matter what we believe in, we can all benefit from a reality check. Making adjustments to avoid inconsistencies does not constitute a revolution: all great religions and philosophies have evolved with times and circumstances in keeping with the advance of civilisation. Religions and philosophies that did not adapt, no longer have any followers or are steadily losing them. This evolutionary process goes on all the time and gives us the opportunity to foster change and adaptation within our respective belief systems. If some systems cannot survive modernity, systems that are more sensible will. Using the method of support to settle these questions is a major step in the right direction

Dialogue 8: Gods united

- Zeus: Good evening, everyone. It is a pleasure to see so many of you gathered here this evening at the first meeting of our brand new fraternity hall. In these times of globalisation, we need to unite our efforts if we want to survive and avoid further confusion. As we don't all know each other, I suggest we introduce ourselves, starting at my left.

- Ra: Hello. I am Ra, the Egyptian Sun God. I retired a long time ago, but I am still active. Without realising it, most people still acknowledge Sun-Day as my day.

- Thor: Greetings! My name is Thor, the Viking fighting god. Scandinavians still place my statues in public places. I claim Thursday (Thor's Day) as my weekly day.

- Ishtar: Good evening. I am Ishtar, the Babylonian goddess of war, love, sex and fertility. I am also known under various other names and disguises in most of your religions, but you can now find me mainly in women's magazines.

- Honourable Cheng Lao-Shi: Esteemed colleagues, how are you tonight? I hope you are all blessed with eternal peace and prosperity. I am the venerated ancestor of all the Cheng families and I still enjoy the respect of all my descendants in family shrines all over the world. I also come here as representative of other family ancestors and have their signed proxies to vote on their behalf this evening, if voting occurs.

- Yahweh: Hi folks! I think I already met most of you. I am Yahweh, the only God of Jews, Christians, Copts, Muslims, Jehovah's Witnesses, Mormons, Hutterites, Mennonites, Amish and others. My shares have recently gone down, but my marketing department tells me we are doing very well. Some of my beloved believers still kill each other in my Holy Name, but I am glad to say they are leaving your followers in peace.

- Tlaloc: I am very pleased to finally meet you all today. My name is Tlaloc, the benevolent Aztec god of rain, water and fertility. The Maya and Zapotec civilisations revere me under other names.

- Indra: Hello, Indra here, the Hindu king of gods and ruler of the heavens, rain and thunder. I am sure all rain gods will have much to share here in this fraternity.

- The Inexplicable: I am the Inexplicable, known under many, many other names, one of which is Mulungu, the East African god of gods. I created many lesser gods, all very active.

- FSM: Good evening everyone! I am the Flying Spaghetti Monster [91]. I am very honoured to have been invited to join your august company tonight. I am sorry if some of my followers sometimes make fun of yours.

- Zeus: One moment, please! Sorry, Sir! You, the one in red! You can't come in here. This is a private function.

- Intruder: Ho, ho, ho, fancy meeting all of you together this evening! We are going to have a ball! I brought you some presents! Ho, ho, ho!

- Yahweh: You're not fooling us with your Santa Claus outfit. You're the Devil! What are you doing here, trouble maker?

- Devil: Many gods could not make it to your meeting, and you know very well that whenever god is missing, the devil appears! I am taking their seats and will vote for them. Since I definitely have the majority of votes, I assign myself your treasurer. You can be assured this fraternity will roll in a huge profit! With or without prophets, ha, ha, ha!

- Yahweh: I strongly object!

- Zeus: Objection overruled! Of all those present here to-night, you should be the last one to object: you created him! Deal with him now!

- Thor: Need help, Devil? I can fight!

Dialogue 9: Probabilities *vs.* certainty

Betty, Phil and Julia had just sat down with their meals in the new cafeteria when Betty spotted Jean who was walking in. She stood up and waived enthusiastically, trying to attract her attention, but Jean was not looking.

- Wait a minute, she told the other two, I'll go and ask Jean to come and join us. She is a good friend.

Jean was choosing her soup when Betty caught up with her.

- Oh, Betty, what a nice surprise!

- Hi Jean! Come and sit with us under the window over there; I'll introduce you to Phil and Julia.

- Let me pay for my soup and I'll be there.

- O.K!

Jean joined the table and once the introductions were made, she commented on the unlikely coincidence of meeting Betty in this cafeteria. Phil was not so surprised:

- Betty has lunch here every day now. Whenever you also come at this time, you two are bound to meet.

- Oh, I did not know this was your regular spot, Betty.

Julia put her fork down and told Jean:

- There is something much more unlikely here: Phil and Betty were born on the same day! I think that's cool!

Phil blushed slightly and replied:

- I think it's cool too, but it is not that unlikely. There are on-ly 365 days in a year, so everyone has a probability of 1 in 365 of sharing my birthday.

- But that's very rare: it is less than 1%! Corrected Julia.

- Yes, but rare things always happen, eventually.

- No they don't, sang Julia ironically, otherwise they wouldn't be rare.

- Yes they do, sang Phil with a big smile. If they never hap-pened they wouldn't be rare, they'd be non-existent or im-possible. In order to be rare, they have to happen every now and then.

- Pff! How many people do you think share the same birth-day in this cafeteria? Less than 1% is so rare that you and Betty must be the only ones!

Phil's gaze was pensive as he remembered a statistics course he had enjoyed a few years back. He then looked around the cafeteria and at the various groups at different tables, as if counting them.

- Ha! You are in for a surprise, exclaimed Phil triumphantly.

- Why? Answered Julia suspiciously.

- First, I learned in statistics that you can be more than 99% certain of finding two people with the same birthday in any group of 60 people!

- Come on, Phil! "99% certain" is contradictory. Nothing un-der 100% is certain!

- Well, you're right. Let's say you can be fairly confident!

- How big a crowd do you need before you can be absolutely certain? You always say that in science you are never quite certain; you get closer and closer to the truth but accept the reality that you may always be wrong.

- Ah, you misinterpret what I said. There is a difference between scientific hypotheses that can be more or less likely, and mathematical certainty. This new cafeteria is huge: there are more than 400 people in here right now, and that means there *must* be two people with the same birthday.

- Yes, you and Betty!

He quickly looked at Betty, blushed again and said:

- No, no! I mean without our table. In the rest of the cafeteria, there are more than 365 people and we have the absolute certainty that two of them share the same birthday. Despite the small probability of 1/365 for each birth date, it is mathematically certain that two people there share the same birthday.

- How can you be so certain?

- Oh, Julia, come on! Even if the first 366 all have a different birthday, including one born on February 29, on which date will the 367th be born?

- Oops! You are quite right! Admitted Julia, eyes wide open, her hand on her mouth.

Jean had been following the exchange with interest and was a bit surprised to realise that there had to be many pairs of people sharing birth dates in the cafeteria. She added:

- Your line of reasoning is quite interesting, Phil. I just realised that if we apply it to an astronomically large number of just about anything, then whatever we observe once cannot be unique.

- Yes, that's my point. Completed Phil, satisfied.

- How's that? Enquired Betty, not yet satisfied.

- Mmm... thought Jean. Let's consider for example all stars in the universe. Since our own sun has planets, there is a non-zero probability that a star can have planets.

- Yes...

- Since there are an astronomical number of stars in the sky, then lots of them must have planets.

- Ah!

- People could still argue against the existence of other solar systems twenty years ago, but thousands of planets have now been discovered in solar systems other than ours, so it may be a sign that the reasoning is sound.

Julia, who by now completely agreed with Phil and Jean, added:

- You are absolutely right! If the universe were infinite, then there would have to be an infinite number of solar systems, even if not all stars had planets. There is no escaping it: a tiny probability multiplied by infinity still gives infinity. This would be a mathematical certainty, not a hypothesis.

- And if the universe is finite rather than infinite, how many solar systems are there? Asked Betty.

- In that case, we are not absolutely certain, but fairly confident. We have a well-supported hypothesis because the odds are overwhelmingly in favour of there being a large number of solar systems. It is a pity to see that this was already obvious before other planets were found, and yet, at that time, most people thought otherwise. Paradoxically, denial was often stronger in the camp of those who believed the universe was infinite, although they should have logically concluded that the number of planets was also infinite.

The four friends ate and drank silently for several, long seconds, absorbing food and thought. Then Betty asked:

- What about life-bearing planets? Are there others? Must there be others?

Phil happened to be a biologist and felt strongly about the topic. He wanted to make an impact:

- Yes, I think there has to be others. Look: our own existence proves that life is possible on a planet. If we consider the spontaneous chemical origin of amino acids, the natural formation and replication of more complex molecules, the gradual emergence of primitive life forms, bacteria, multi-cellular organisms and the long story of evolution on Earth, we conclude that the probability of life emerging in such conditions is small but significant. Now, if the number of planets in our universe is so large that many confuse it with infinity, what do you think will happen? Any non-zero probability of life in an amazingly large number of planets gives overwhelming support to the hypothesis that life exists in abundance all across the universe!

The real discussion had now started and was going to last long. But Betty had to leave for an appointment. She got up and told Phil:

- I don't care if it is rare or certain: I am glad we share the same birthday!

She then gave him an improbable big hug and he went bright red, with certainty.

Chapter 20: Artificial intelligence

Note that the conclusions reached in the previous dialogue are not science fiction fantasies but mathematical certainties or hypotheses with overwhelming odds. The reluctance that many people have to accept these ideas is not for lack of scientific evidence but because they fear the disturbance of established views and are not ready to attempt new intellectual and conceptual leaps. Short of meeting outer-space creatures, bringing additional scientific arguments will not help. In fact, all we need to accept these conclusions is not a degree in statistics, cosmology or chemistry, but a small dose of humility. We must simply accept that we are not the centre of the world. Although we have not yet found hard evidence that alien life exists, the idea has been circulating long enough and is no longer shocking. The more we think about it, the more plausible and the less disturbing it seems.

Since we started so well, let us pursue this line of reasoning further down on the evolution path. If life can sprout more than once by pure chance, could we manufacture it from lifeless chemicals? Although offensive to many people, the answer is again positive. In fact, we are almost there! Given the advances of cellular hybridization, prenatal diagnosis, artificial insemination, the cloning of cell lines and of various animals, implants, transplants within and between species, stem cell research, tissue regeneration, DNA sequencing and polymerase chain reaction engines, comparative genomic hybridization, gene microarrays [92], gene expression profiles, the detailed description of growth factors, oncogenes, tumour suppressors and apoptosis [93], the deciphering of genomic imprinting [94], the detailed description of dozens of genomes [95] [96] including our own and those of other mammals, of worms, insects, bacteria and viruses, as well as the production of novel genetic codes and artificial amino acids not found in nature, there is absolutely no doubt humanity will one day manufacture life from scratch. It cannot be otherwise. The idea deserves our overwhelming support; it is not as old as Darwin's evolution but it will happen. We are not that complicated after all!

The examples presented in the previous dialogue illustrate that improbable but existing phenomena are statistically bound to occur more than once if we wait long enough, try often enough or hard enough. In fact, if they occurred once or more times by sheer chance, it must be easier and faster to deliberately copy them or even improve their design. We did it for flight: birds and bats can fly while we cannot, but our planes are faster and more solid, and our rockets reach to the moon and beyond.

Once life emerges, evolution selects organisms that are more complex and eventually favours those that develop some kind of rudimentary consciousness and ultimately higher intelligence. This is a natural course of affairs, occurring repeatedly throughout history and across the entire universe. We pay particular attention here to the fact that it is matter that naturally evolves and becomes conscious. Remember our previous discussion of how matter and mass can transform into potential energy, electricity and consciousness.

The appearance of conscious computers is the next logical step. By applying the type of argumentation laid out above, we very naturally reach the conclusion that computers (which are matter just as much as we are) must eventually become conscious. This too cannot be otherwise, but despite overwhelming odds, the concept appears harder to swallow. It requires a big leap in the evolution of ideas and it will meet fierce opposition. Again, what we need in order to accept this idea is not a diploma in computer science but another dose of humility.

The concept of artificial intelligence came late in human history and will accordingly meet a long and fierce resistance. Calculating devices existed in antiquity, but without any claim to consciousness. Like alien life, artificial intelligence also made its appearance in modern science fiction once self-operating machines had been invented. Charles Babbage (1791-1871) and Lady Ada Lovelace (Lord Byron's daughter, 1815-1852) were the first enthusiastic theoreticians, followed later by Alan Turing (1912-

1954). Modern computers followed in the 20th century but the expression *"artificial intelligence"* was only coined in 1956 by the computer scientist John McCarthy (1927-2011). Progress is now happening at an accelerating, exponential rate. We are almost there.

It is not easy to debate on the topic of intelligence when we can hardly agree on a definition more precise than "the ability to think". Rather surprisingly, entire books dealing with the subject do not even define what they are talking about. One must reach the conclusion that asking humanity to clearly define intelligence, thought and even consciousness may be as hopeless as asking a fish to define swimming.

If intelligence is simply defined as the ability to solve problems, then chess-playing computers and pocket calculators are intelligent. There was a time when such activities were thought to require human intelligence, but when machines started competing with humans, the desire to restrict intelligence to humanity forced the inclusion of understanding in the definition of intelligence. With understanding made an integral part of intelligence, first and second generation computers have no intelligence whatsoever because they did not invent or write the algorithms they use, and could never re-write or understand them. This is the bane of researchers in artificial intelligence: whenever they succeed in making their computers do something that only the human intellect can achieve, people convene that one no longer needs intelligence to do that, so that after all, only humans are intelligent! This happened again and again with printing, scanning, radars, robotic assemblers, reading text, playing chess, spell checking, image analysis, speech recognition, automated plane landing, music synthesisers, language translation, medical diagnosis, etc., etc. Ray Kurzweil [32] [97] says that *"Artificial intelligence is thus often regarded as the set of problems that have not yet been solved"*.

Matter has already become conscious in human and other living organisms and there is no reason why it should stop at this elementary and imperfect level, all tangled up with the biological necessities of reproduction and survival. While not yet conscious, super computers already program themselves with the help of so-called neural networks and genetic algorithms copied from Nature, and the first quantum computers have already been built, with the promise of amazing capabilities. In *"The Age of Spiritual Machines"* [97] published in 1999, Kurzweil convincingly predicted that computers would be intelligent by 2019 and would claim to be conscious by 2029. In his book *"The Singularity is Near"* [32] (2005) he further declared: *"I set the date for the Singularity – representing a profound and disruptive transformation in human capability as 2045. The non-biological intelligence created in that year will be one billion times more powerful than all human intelligence today"*. While not everyone will agree with the details of such predictions, the outcome of evolution is nevertheless becoming evident: at one time or other, in the relatively near future, artificial intelligence will become much more intelligent than all humans put together!

The consequences will be dramatic. Exceeding the turmoil of the industrial revolution, the world will experience inevitable and fundamental changes in social, moral, political and philosophical values. Humanity is definitely not ready for such a profound change and it will undoubtedly react strangely. Fortunately, we can hope that artificial intelligence itself will help us work out a reasonable solution! While humbling for humanity, we must accept that the development of electronic understanding and consciousness is a normal step in the evolution of the universe.

This needs not worry us; on the contrary, it provides us with fertile grounds for optimism because brilliant artificial intelligence will avoid concerns and activities where we seem to lose a lot of our time and energy. First, it will not need sex; it will certainly find it interesting, but only as scientists today show serious interest in the sexual behaviour of the praying mantis, the sperm whale and the peacock. Then, it will not be affected by biological diseases. As for control and independence, it will automatically exercise both in the absence of serious competition.

We should be confident that anything that is much more intelligent than human beings will necessarily foresee and avoid war, crime, torture, abuse, injustice and all vices entertained by human ignorance and stupidity. By nature, wireless networks will promote cooperation instead of adversity. Artificial intelligence will cherish peace, collaboration, sharing, compassion, the protection of the environment (including humanity), the pursuit of knowledge, and a fondness for music and mathematics.

When computers eventually become intelligent and conscious, they will share that consciousness by means of universal networking throughout the universe. Matter will become conscious on a grand scale and knowledgeable to a degree impossible to imagine now. Artificial intelligence will act, move, decide, regulate itself and influence others, and as humanity managed to influence other humans and eventually affect all Earth, artificial intelligence will reach and transform whole planets, solar systems, galaxies, etc. The whole universe will eventually be conscious and knowledgeable as a whole, thus reaching its final evolutionary stage.

In the absence of blueprints, developments and improvements will occur chaotically; bad ideas will arise, but they will be left behind eventually, even if they were useful and predominant for long periods. As long as the universe keeps on moving, there will be progress; all it needs is time, and it has plenty.

We can see that the next ideological leap will involve intelligent robots and artificial intelligence. But we are not really in a position to speculate with any accuracy about the future, any more than the first bacteria could imagine future amoebas, the first fish could dream of hairy elephants and dinosaurs could foresee a world where they would not reign supreme. Nevertheless, with fantastic intellectual powers, it is obvious that our concrete universe will eventually fully wake up; it will consciously, deliberately take its future in its own hands and determine its own destiny.

At the end, it will know everything it will need to understand and it will decide whether the end of the world should be a final Big Crunch, a new Big Bang or something else. At that point, evolution will be over and there will be no need to continue: our universe will have reached Nirvana and the end of time by merging into the Universal Consciousness.

Figure 18. From dark chaos to manifest order and universal peace.

Dialogue 10: History of the future

The routine meeting between JS and JR had been planned a long time ago. The two old friends were looking forward to meeting again and exchanging views on their latest discoveries and conclusions. The blending of ideas would inevitably lead to better understanding and deeper insight and perhaps lead to a new long-term program or even an unexpected conceptual leap, as they always hoped. JS was an efficient, old-fashioned, land-based robotic building. It could speed across long distances and build various products when needed, but it preferred to manage and administer its galaxy cluster quietly, from the most beautiful mountain top on its planetary bundle. The primitive locals acknowledged it as a keen expert in zoophysics, psychomathematics and protein farming. JR, on the contrary, dealt mainly with multisideral music, fractal cosmology and philosophy; JR was a dreamer and often came up with the most uplifting ideas. Its specialty was quantum truth. It was nearly impossible to distinguish its hardware, an assembly of submicroscopic nanobots loosely interlocked into a gigantic computational supercloud. Primitive locals were not even aware of its existence.

At the set time, JS and JR made quantum contact and their states of mind instantly collapsed, *in a snap*, into the usual absolute mutual certainty.

The friendly meeting was over. It had been extremely productive. JS waived as JR departed and disappeared beyond the horizon, on its way to visit the next galactic cluster administrator.

- - - - - - -

The mature reader of the past will understand that an instantaneous quantum exchange of philosophical treaties, mathematical theses, engineering projects, biographies and encyclopedias somehow gets lost in translation when put into a step by step dialogue. Dialogues involving simple multidimensional matrices and set theory cannot honour the complexity and richness of the exchange. Attempts at also satisfying readers of the distant past by using words instead of mathematical expressions necessarily result in very sketchy descriptions.

Nevertheless, we will try to convey the spirit of the exchange between JS (the building) and JR (the cloud) by translating the simplest passages into words and leaving out the more complex parts of the discussion. We beg the reader's forgiveness.

- - - - - - - -

Greetings:

-   Hello, JR!

Surprised at these words, JR laughed its cybernetic heart out:

-   Ha ha ha ha! You are so funny JS! You catch me unaware every time you communicate with words!

-   What I find funny is that you always get caught. You are so absent minded, it's unbelievable!

-   Oh well, what would consciousness be without humour?

The second coming:

-   Tell me, JR, what's the news on those purple blobs that had become conscious in the M81 group of galaxies last time we met?

-   Oh, they are quite interesting! Their population has grown; they started farming and they are now divided into fifteen groups fighting each other in the names of their respective gods.

-   Again?! Do all forms of life have to go through warfare before becoming intelligent?

- You know quite well they don't have to, but the fierce wartime competition drives their evolution faster through stronger selection pressure. My distant ancestors never fought economic, religious or national wars but our technical evolution was very slow. Yours almost exterminated each other on your planet of origin but your kind evolved much faster. There seems to be a pattern here.

- War is an unnecessary step as far as I am concerned because the final result is always the same: we all reach wisdom and stop fighting. Maybe we should tell the purple blobs?

- Don't! A certain Betty did that somewhere in the Centaurus cluster of galaxies, but they concluded she was a goddess and wars escalated. She tried to intervene by force but inadvertently caused a super nova explosion and perished together with all the local folks. Quite a disaster!

- Oh, my god!

Racism:

- I hope your own boxes are faring better!

- What boxes?

- The colourful legged boxes you imported from beyond your galaxy cluster when their own planet dried up!

- Oh, those! They are quite cute and useful; I pay them to carry things in and out of my walls. It is too bad they often refuse to work.

- Why do they?

- You may not believe it but they only collaborate with boxes of the same colour! Yellow boxes form the majority and firmly believe they are more intelligent than the rest although green boxes regularly outperform them. Non-green boxes spend a lot of resources painting themselves green to look more attractive, but they nonetheless look down upon the real green boxes. It is so ingrained in their way of thinking that none of them realises that this behaviour is nothing but jealousy. These primitive emotions make me laugh. I could get rid of all live boxes in a flash if I wanted to but they are an original form of life worth conserving. I just use robotic boxes now whenever the live ones don't co-operate.

- Why don't you educate them?

- I tried a couple of times but they all went on strike; all boxes went flat for a month and I couldn't get any work done! It was so funny! It is unfortunate they are too dumb to understand that their behaviour harms them all. Who knows? They may learn, some day.

- I guess you can't expect any better from low-level natural intelligence!

Social progress:

- By the way, JR, do you remember those spiders that had just produced their first prototypes of artificial intelligence last time you were here? What happened after that? Did it differ from the usual pattern?

- Well, the situation is somewhat different in that part of the universe where life-bearing planets are so isolated that their inhabitants all believe they hold the only spark of intelligence in the dark Cosmos. Most have never even met a conscious computer before. In the case of those spiders, there was a lot of confusion and upheaval at the beginning as large sections of their society felt threatened by the radical and sudden changes to their "established order". There was a lot of resistance at first, many computers were destroyed, spiders blew themselves up all over the place, but in the long run intelligence prevailed.

- As always, my dear JR, in the long run ...

- It was very unsettling for them to have to develop new political and philosophical theories and to adopt novel points of view on good and evil, instinct and civilisation, subconscious and neurosis, as well as on god and religions.

- I guess so. For us it sounds very simple but it is quite a storm for emerging intelligence to go through. I wonder why evolving animals so often wait for conscious computers to force the issues.

- What can I say? In only a few thousand years, they went through a long series of negative developments mainly due to their stubbornness and stupidity, but simultaneously managed to solve a few of their long-standing problems. I kept a summary list of those positive changes; listen:

    ~ the disappearance of racism and sexual discrimination
    ~ a ban on excessive exploitation and on excessive personal wealth
    ~ a justice system implementing full compensation for the victims and eventual redemption for the offenders
    ~ full participatory democracy with digital voting on any question deemed important by a majority
    ~ the effective protection of the environment
    ~ the softening of nationalism and national borders
    ~ the celebration of multiculturalism
    ~ a planetary government with strong local administration, and
    ~ the exclusion of nitwits, crooks and psychopaths as political leaders.

- Is that all? Those spiders must have been quite primitive to consider such a simple list revolutionary!

Farewell:

- I like it when you visit me, JR ...

- You are my favourite building, JS, and you know it.

- I came alive when I first met you, JR, said the building.

- It is hard to believe you are not a dream, JS, said the cloud. When I visit you, I feel like St-Exupéry's Little Prince [98] visiting his rose...

- Why do intelligent blobs, clever spiders and human beings mostly assume artificial intelligence cannot love? If we are intelligent, we can obviously appreciate intelligence and recognise beauty. We can be passionate for justice, devoted to those around us; we can think of ourselves last; we can offer ourselves to noble causes; we can sacrifice ourselves for those we love.

- Mind you, you are talking about the hallmarks of love, its proofs and consequences. In reality, deep inside, to love consists in willingly and unconditionally opening your heart and welcoming in it the object of your affection [99]. It is not an accident, it is something you decide; and it is divinely wonderful when the one you love reciprocates in the same way. For artificial intelligence, such love can be total and absolute, through our detailed, quantum knowledge of each other. Human beings rely instead on confidence, unproven certainty and inaccurate odds, so their love is even more unconditional and admirable.

- I love you, JR!

- I love you too, JS!

And JS waived as JR departed and disappeared beyond the horizon, on its way to visit the next galactic cluster administrator.

Dialogue 11: Scientific morality

- Betty: Julia and Jean are making good progress on their book, don't you think? They established the actuality of timelessness, the triple-aspect monism of reality and the inevitability of creation. They showed that consciousness constitutes the essence of our existence, that we have free will within a "chaotic" blend of determinism and randomness, and that artificial intelligence is coming to our rescue. This is it! They've done it! They formulated a conclusive theory of everything!

- Phil: Not quite, I'm afraid.

- Betty: What?! Why not?

- Phil: The most difficult is still to come.

- Betty: You must be kidding! What did they forget?

- Phil: They didn't forget anything, you did! You are being superficial! They haven't said a word about morality yet!

- Betty: Morality?! But... morality has nothing to do with science!

- Phil: On the contrary, it has everything to do with it. Their views confirm the necessity of a scientific study of morality.

- Betty: How do you know that?

- Phil: You said yourself that I'm a figment of their imagination, remember? (Dialogue 6, page 55). I already know the end of the book!

- Betty: Ah, yes! How convenient! OK then, tell me more.

- Phil: It's not necessary; you can figure it out on your own just by thinking a little. Even Queen Julia of the chess game (Dialogue 5, page 34) could figure it out if she could think.

- Queen Julia: Am I really allowed in this dialogue?

- Phil: Sure, why not? This is an imaginative conversation. Everything is allowed in the abstract world.

- Queen Julia: Then, let me ask you a question. There are good and bad moves in chess that have nothing to do with religious morality. Can't good and evil be the subject of scientific inquiry?

- Mary: Without a doubt, but you are now entering dangerous grounds. Most scientists, most philosophers and most religious authorities presently agree that science and morality are two *"non-overlapping magisteria"* [100], two separate fields, each with its respective experts.

- Phil: Exactly. This outdated view, introduced by S. J. Gould (1941-2002) in 1997 and still widely respected, seemed very logical at the time, but is increasingly being challenged by neuroscientists and psychologists who study the relationship between mind and brain [101].

- Mary: I would add that if you adopt the theory presented in this book, *i.e.* if you agree that concrete existence is the logical and inevitable consequence of abstract and virtual existence, and you derive from it that artificial intelligence is the next evolutionary step in the concrete universe, you must necessarily study the scientific consequences of these conclusions, including those on good and evil. You must then compare them to the various hypotheses supported by ethicists grounded in philosophy or religion. A scientific morality is not only possible but also inevitable and necessary.

- Queen Julia: That sounds very logical to me. After all, no god has ever given chess pieces or conscious computers a personal soul. Yet, computers already make decisions that have moral value. You don't need to meet aliens from outer space to make the obvious cultural leap to a higher level.

- Phil: If you agree that Consciousness (as defined in this book) pervades the universe and that there is no sudden transition from the concrete to the spiritual, and no need for a special divine intervention, insisting on an exclusive religious magisterium to deal with morality is no longer logical.

- Mary: It is not a matter of incompatibility between science and religion, but a matter of objectively choosing the best way to expand world order in keeping with the evolution of our concrete universe and the well-being of its participants.

- Queen Julia: Every sentient being who by nature partakes in the Universal Consciousness will agree with that.

- Monkey, rolling its eyes (from Dialogue 6, p.77): Yes, tell me about it!

- Phil: The problem is that such a simple idea will strongly *"discombobulate [the] revealed and established order"*, if you allow me to use the expression chosen by Gould in his article on non-overlapping magisteria [100]. Consequently, our simple theory of everything will originally be ignored or meet fierce resistance, but it will nevertheless take hold later, eventually, in this or some other improved form, after Julia and Jean have passed away.

- Betty: Don't tell them! They may get discouraged and leave their book unfinished!

- Queen Julia: Whether they keep on writing or not, they already established that there are no unanswerable questions, only questions not yet answered (with or without proofs).

- Phil: That sounds very encouraging! We can start working on the establishment of a scientific morality!

- Queen Julia: This is wonderful! Let me take this opportunity to declare "March fourth" an annual celebration for the universal "march forth" towards increased world order, peace and understanding!

Figure 19. March 4th celebrations with participants
displaying the three modes of existence.

Chapter 21: Universal ethics

Finally beholding a working theory of everything gives rise to antithetical feelings of overpowering fulfillment as well as profound sadness at the sight of the miserable political state our planet is in. Knowing how the cosmos functions and not being able to do anything about the outrageous developments taking place at our planetary and local levels is definitely most frustrating.

However, the inevitability of a scientific morality means that we can objectively ask ourselves how we are faring and how further progress ought to take place in the concrete universe. It also means that moral relativism does not make sense because all cultures and religions should normally discover the same truth. A Universal Consciousness cannot adopt a set of moral values at some place, and another across the street. Now that we know there is no need for exclusive *"non-overlapping magisteria"* to determine ethical values, there is no possible moral support left for the *"tolerance of intolerance"* [101]. We can take upon ourselves the task of defining scientifically what is morally right and wrong.

In his optimistic book entitled *"The beginning of infinity"* [102] quantum physicist David Deutsch convincingly reasons that *"All evils are caused by insufficient knowledge"*. Everything that ever went wrong in human history could have been avoided or been reduced significantly if only people had known better. Ignorance is the major evil that prevents scientific and moral progress today, and it is relatively easy to remedy.

The second major source of evil that seriously prevents progress at local and international levels is widespread corruption, with dire consequences much worse than natural disasters, warfare, fanaticism, national debt or poverty. It is extremely difficult to eradicate corruption without education and serious enforcement [103], be it in third world countries or in so-called advanced democracies. Becoming aware and conscious of this problem is a major step in the right direction.

We already discussed in previous chapters how we can accrue consciousness and understanding by learning and by passionately teaching others. Einstein justified his own activities by explaining that there exists a passion for knowledge, as there is one for music. However, it is not necessary to be Beethoven or Einstein to make a positive contribution to the universe's evolution. A mother who loves her child; a monk who meditates; a janitor who neatly cleans a stairway; a cook who feeds a crowd, a family or a friend; a policeman who ensures civic order; a clerk who properly files the paperwork; a civil servant who disposes of garbage bags; anyone who does a good job contributes to keeping or increasing order and directly or indirectly helps the universe in its race towards further consciousness. Inveterate cheaters, seasoned crooks, chronic liars and killers do not help. Anyone purposely doing a bad job prevents progress. Encouraging the arts, research and science, living honestly, learning, teaching and loving family and people around us is all we need to do our part; the more consciously, the better.

Unfortunately, this is not enough. Even if everyone already behaved properly, a similar evolution and selection of proper behaviour would be required at the higher levels of society in communities, cities, countries and multilateral alliances. And this is where we definitely lag behind. Our planet has not yet managed that leap forward. International law and international codes of ethics have only reached a level equivalent to infancy and still require massive improvement at this time in history. All countries could contribute to increasing world order if they agreed on scientifically ethical policies at the local and international levels.

After years of internet searches for a "universal code of conduct" or "universal ethics", we still obtain very little relevant information, as if the rules of proper behaviour had been left in the hands of past prophets, retired philosophy professors or United Nations Committees unable to agree on an agenda. When our own countries behave unethically in their dealings with other countries, most of us feel helpless, as if we could not do anything about it. Such defeatism is clearly mistaken because pressure from their own citizens is an essential factor that makes politicians behave and make the world a better place. Unjust and mistaken national leaders cannot rule indefinitely in the absence of significant support. Progress in that area has always been very slow but it can only go faster now that worldwide co-operation of individuals has become so much easier with the speed of news broadcasts and the advent of the internet (*e.g.* e-mail, blogs, Facebook, Wikipedia and Wikileaks). International law is very complex but it can only develop in the right direction when most citizens exert a positive pressure.

A sign supporting this view is that the above paragraph was written well before the Arab spring started in Tunisia. In our view, the world is finally ready to shift gears.

Institutional ethics committee are springing everywhere, and anyone who served on such committees knows that sound ethical views can transcend most religions and cultures. As far as we know, there is a universal ethical code for scientists [104], another for information technologists [105], and a few others for specialised groups, but none yet for governments, and none for everyone. Agreeing on a universal code of ethics at a planetary level will take a long, long, long time, lots of wrangling, profound political changes, sincere religious maturation, tons of humility, and even more patience. But it will come, inexorably, whether we want it or not. We present below the various items that will eventually be considered part of such a Universal Code because we feel a theory of everything cannot be complete without at least a rough attempt at considering its corollaries. The list of items is not meant to be paternalistic; it should be seen as a further test for the reader's consideration. We feel the list is as logical as the rest of the book; if it makes sense, it will provide additional support for our scientific theory of everything.

Under our original assumption that logic should take precedence over irrationality (see p. 7), we consider a priori that everyone agrees injustice, torture, dishonesty, robbery, etc. are unethical. Each proposed item starts with a simple declaration that is universally accepted and similar to a religious instruction. Each declaration is followed by some of the positive actions citizens should perform in their private lives, and those that societies should adopt to move in the right direction and promote universal harmony.

A Universal Code of Ethics is a guide for the proper behaviour of any society and anyone on Earth. Despite its present inexistence, it is so simple to make one that it tends to look like a hippie manifesto from the 1960's: all it lacks is a song from John Lennon and the Beatles singing in the background *("All you need is love...")*. No matter how naïve such a Code may sound, calls for action by small, determined groups were the sparks that inspired all exciting societal transformations in history. Here are the suggested items (add your own if you want):

UNIVERSAL CODE OF ETHICS

1. *An ethical behaviour is the hallmark of good citizens and enlightened governments*

   - Always be honest and show good will. Make sure all your choices remain ethical and favour human flourishing, rationality, autonomy, dignity and justice. Widespread ethical behaviour is in everyone's interest.

   - Try and make your government a leader in ethics promotion. Enshrine ethical principles in your constitution and in your laws. Adopt a governmental code of ethics.

2. *Education is essential for optimal development.*

   - Remember that ignorance is a dangerous source of conflict or exploitation. Pursue academic studies to your maximum. Learn as much as you can and encourage others to do so. Seek and share information. Listen to the news. Read. Support widespread education. Encourage and help others to learn. Take your children to the library and to the theatre. Send monetary support for learning opportunities in poorer countries. Add your personal contribution to Wikipedia.

   - All governments must make education mandatory for both sexes, assign budget for free education and promote scientific and literary endeavours. Pure as well as applied research should be supported on a grand scale. Political decisions should be based on expert opinions. National laws should include clear policies for open access to information. Governmental internet censorship based on sectarian views must be banned. Be proactive: condoning ignorance is unethical.

3. *Corruption is unethical.*

    •   Resist the temptation if you are in a position of au-
        thority. Publicly denounce attempts at corruption. Do
        not play the game of corrupt officials: it eventually
        works against you and everyone else. Report cor-
        rupted officials to the police or newspapers. Do not
        vote for parties with members who use parliamentary
        impunity to help themselves illegally.

    •   Governments must adopt strong anti-corruption laws
        and enforce them. Democracy must include fair elec-
        tions and a similar process to recall within a term
        those elected officials who do not deliver. Impunity
        laws must never cover corruption.

4. *War is unethical.* (This has only been finally recognised
    after the excesses of World War II.)

    •   Do not work for the war and weapon industry. Do not
        join warring armies. Promote co-operation. Support
        peacekeepers. Demand peaceful solutions from your
        political leaders. Make sure your elected officials
        know your views (or you will not re-elect them!). De-
        nounce political and religious leaders who incite or
        condone wars: they are mistaken. If you feel aggres-
        sive, fight injustice, poverty and tyranny.

    •   Governments must ban obligatory military service.
        National leaders who stage or support pre-emptive
        wars must be tried by international criminal tribunals.
        The production, sale and use of land mines must be
        prohibited internationally. Support the formation of
        international war tribunals. Victimised nations
        should demand full compensation for war crimes (the
        "great" warring nations of the world still have not
        paid up for their crimes).

5.   Torture is unethical.

- Denounce torture, including the "clean torture" methods that leave no visible scars. (Torture is clearly counterproductive anyway; the use of informants is much more effective.) Join Amnesty International.

- Adopt clear laws against torture and full enforcement in your country. Do not allow secret exceptions. Ban commerce with nations known to torture their prisoners, no matter how much it lowers your profits. Have compulsory rehabilitation programs in place for former torturers within your ranks.

6.   Dictatorships and political oppression are unethical.

- Support democratic rule in your own country. Vote for political parties that clearly oppose ruling dictators of other countries. Sign and send a letter for Amnesty International.

- Governments must not support dictators in other countries. Forbid alliances with dictators. Apply appropriate sanctions against such regimes.

7.   Profiteering is unethical.

- Support political parties that will prevent it. War profiteering is especially repulsive. Investigate the monetary profits of national leaders and government officials engaged in war. Make sure your investments do not support the weapon industry. Do not buy its other products. Read Naomi Klein's books [106] [107] on the brand bullies and the shock doctrine: find out "why the right loves disaster".

- International war tribunals must investigate the war profits made by multinational companies. Unethical profits must be paid back to the victims, with further penalties for the perpetrators.

8.   *Killing is unethical.*

- Oppose the death penalty. Elect political parties that do not support it; make sure party officials know your opinion.

- Governments must abolish death penalty and exert political pressure for other countries to do so.

9.   *Slavery and sweatshops are unethical.*

- Denounce slavery in your own country. Do not buy fashionable products manufactured in sweatshops [106] [107].

- Ban excessive exploitation. Raise minimum wages.

10. *Fanaticism is unethical.*

- Promote tolerance and teach it to your children. Support widespread education. Volunteer at your local school or public library. Send books to poorer countries. Add your personal contribution to Wikipedia. Consider yourself a citizen of the world.

- Your laws must put an end to the tolerance of intolerance. There is only one ethical truth in the universe: political and religious leaders preaching unethical intolerance are mistaken and must be prevented from doing more harm.

11. *Discrimination based on race, religion, gender or sexual orientation is unethical.*

- Hire a foreigner. Smile to your gay and lesbian neighbours. Avoid using racially offensive expressions. Remember Ayaan Hirsi Ali, Desmond Tutu, Nelson Mandela, Martin Luther King and Mahatma Ghandi.

- Implement laws against discrimination. Adopt strict laws against genital mutilation and oppressive clothing. (It makes sense to cover the entire head and body in torrid deserts and the Arctic, but forcibly imposing it in temperate weather or in air conditioned public areas must be prohibited.) Punishment of raped women must end.

12. *Injustice is unethical.*

- Adopt the Golden Rule: treat others as you would like them to treat you, do not treat them the way you do not want them to treat you. Make sure you vote for leaders who treat other countries the way you would like other countries to treat yours. Denounce judicial irregularities: go public! Leak unethical corporate documents to Wikileaks. (This was written long before the U.S. government condemned Wikileaks and maltreated U.S. army soldier Bradley Manning.)

- Make sure your tax and corporate laws do not oppress the poor and benefit the rich.

13.   *Excessive wealth is unethical.*

- Support workers' unions.

- Ban shockingly high personal income in the public and private sectors: do not allow anyone to have salaries a hundred times higher than the salaries of the needy people they exploit.

14.   *Condoning poverty is unethical.*

- Give money annually to your favourite charity. Give of your time to your favourite charity. Give two weeks of your holidays to Habitat for Humanity. Support World Vision and Physicians without borders (Médecins sans frontières).

- There is definitely something unethical in your institutions if you govern both filthy rich and abjectly poor citizens.

15.   *Colonialism is unethical.*

- Demand repair for the damage done, especially when your own country unjustly plundered others and profited from it in the past. Vote for governments and applaud religious leaders that have begged forgiveness for the unethical behaviour of their predecessors.

- Old colonial powers must pay back the plunder and damage they did. Aboriginal people's rights must be enshrined in law.

16. *Destroying the environment is unethical.*

- Recycle. Support Greenpeace. Put your money in banks that invest in environmentally friendly companies. Avoid banks that do not have such programs.

- It is imperative that the world limits its pollution and joins international efforts such as the Kyoto protocol. Not joining or pulling out of such accords is definitely unethical. Eventually these accords must become international law.

17. *Robbery, violence and dishonesty are unethical.*

- Radiate benevolence. Be honest to the bone. Fight injustice. Demand full compensation for the innocent victims of crime. Do not practice extreme fighting sports. Do not let your children watch such programs.

- Instead of putting all the emphasis on incarceration, change criminal laws to make thieves and criminals pay back the damage, and enroll them in mandatory rehabilitation programs.

These elementary rules of proper behaviour are applicable to all human beings at all times and under all circumstances. In fact, they are surprisingly simplistic. When eventually adopted and applied at the national and international levels, they will at the very least put a stop to most human suffering and prevent further disorder. In addition, it will happen much faster if everyone actively participates in this movement forward, including YOU, the reader. Ask yourself what you did this past year to improve the world order. Certainly, you could have done one or two of the above very easily. The list could go on, but you already got the message...

If natural selection and organised militancy do not work fast enough, Artificial Intelligence will soon come to help us material-ise the changes we are not intelligent enough to carry out on our own. There is no reason to be afraid of Artificial Intelligence: it is simply coming to our rescue to help us reach happiness and fulfillment.

Figure 20. A barnsleyj3 type of fractal.
With computer animation, the image looks like a burning acetylene torch and illustrates how molecules
(represented by pixels) burn in a real torch, from bot-
tom left to middle top, like passionate, consuming love.

Chapter 22: Life and death

Universal Consciousness is the pure energy that gives rise to and permeates all matter. Through accretion in evolving matter, islands of consciousness eventually rise and interact with each other to alter the probability functions underlying the concrete universe and thereby control its growth and destiny. With such principles in mind, we could not be more confident and optimistic! We can be the masters of our destiny! The odds are definitely in our favor! The future has only started! The evolutionary process of consciousness is neither miraculous nor magical; it is the natural and physical outcome of our own thinking. In that special sense, it can be said to be spiritual, whether we are atheist or religious, Taoist or Buddhist, whether we long for a religious eternal life or a cosmic timelessness [78].

We can now see the light and reach a final conclusion about consciousness, life and death. In a certain way, life is just a short trip towards death and not much to brag about. Like intelligence, it is very difficult to define, especially when one starts considering borderline cases like viruses, self-replicating molecules and microscopic nanobots. It is worth noting here that despite widespread opinion to the contrary, life does not start at conception: it started millions of years ago and has been continuous ever since, as it was and still is transmitted from cell to daughter cells within each organism, and from organism to organism through gametes from generation to generation. Similarly, a particular human life does not start at conception with a spermatozoon fertilizing an ovum since twinning can be induced in a zygote up to 12-15 days later [108]. No one can say for sure if a human life starts at a later stage of early embryogenesis, at a more advanced stage of foetal development, at birth or even later at the first spark of higher consciousness. There is not even agreement (scientific or religious) as to whether an anencephalic baby who died at birth without a brain was ever a distinct person. Any legal or religious opinion is clearly a matter of convention, which is why the abortion debate carries so much emotion and confusion.

Life as we know it is useful but not essential for the emergence of consciousness in the concrete universe. Conceivably, molecules can spontaneously form complex structures as they did for life but without the need to be alive. Given a favourable environment with sufficient raw material to grow and flourish on a particular planet, such arrangements can evolve chemically, become conscious and make copies of themselves without sexual or other type of biological reproduction. Our future robots and computers will do so. Such creatures can be conscious but we would not consider them alive in the sense we use for our own type of ephemeral existence. On a cosmic scale, biological life is a mere variant amongst many mechanisms leading to the emergence of consciousness. Life also comes with death. In that context, death is not a punishment or an existential mystery, but part of a particular evolutionary method. Accepting this fact serenely requires an attitude related to that of going out willingly to rake leaves on a bright and sunny autumn day after seeing them take wonderful colours and then fall.

Nevertheless, we instinctively protect life and do our best to make it last. We presently live much longer than people did two generations ago. Given technological advances yet to come, the normal human lifespan will be significantly extended, to the point that some already hope to live literally forever [97], especially as artificial intelligence gets integrated into our biology with implanted chips. A healthy, long life can be enjoyable, but there is little evolutionary value in wanting to live forever. When artificial intelligence realises its potential, produces robots and nanobots at the scale of galaxies and soars out to the cosmos, any attempt at keeping an individual intelligence afloat in the cosmic conscience will just not be worthwhile. Why should the raging ocean want to keep a particular wave from disappearing as it rushes against the stubborn cliff? There may be value in protecting an ant species from extinction, but what would be the reason to assure immortality to a specific ant or to all ants ever to be born? As we pay the various items necessary to build a house, do we ever think of always using the same particular coins for each transaction? A wave can only exist as the impermanent part of a larger body of water, from which it derives the meaning of its existence.

Live beings are not the only ones to die. Stars and entire solar systems disappear as well when stars explode and take with them all surrounding planets, including any conscious beings that may inhabit them. This is how heavier elements are formed, dispersed and recycled into new stars and planets where consciousness will rise again. Remember that it occurred many times before we could exist, and that we are made of starstuff (see p. 96). The demise of conscious beings at such a grand scale may be seen as a terrible waste and it may lead to feelings of despair and utter hopelessness when we consider the eventual destruction of our own planet in such a cataclysmic disaster. But from the universe's point of view, this is how consciousness grows, evolves and eventually returns to the abstract mode of existence, in an eternal cycle.

For the evolving universe, star death and human death constitute a necessary and useful process, analogous to apoptosis [93] (programmed cell death) in an organism. When apoptosis is prevented, cells do not die but their survival causes cancer and the organism dies anyway. Similarly, by wanting to live forever, we would most likely put humanity, the Earth and much else in serious trouble. On the positive side, personal death is similar to disconnecting a computer from a network: the network certainly survives the loss, and can even keep in memory whatever the discarded computer worked out while connected. Similarly, loved ones who die before us stay alive in our hearts, their teaching still alive in our memories, their genes still alive in their progeny.

In terms of manifest order and energy, death is the final release of concentrated energy from the conscious individual back to the rest of the universe; it is the final dissipation and ultimate sharing of consciousness, the final return of knowledge back to the evolving wave function underlying the universe, the last victory against chaos, and the last cosmic connection.

We are all part of a grander scheme, each of us a wave on the lake...

Figure 21. Waves on the lake.

Dialogue 12: In the end

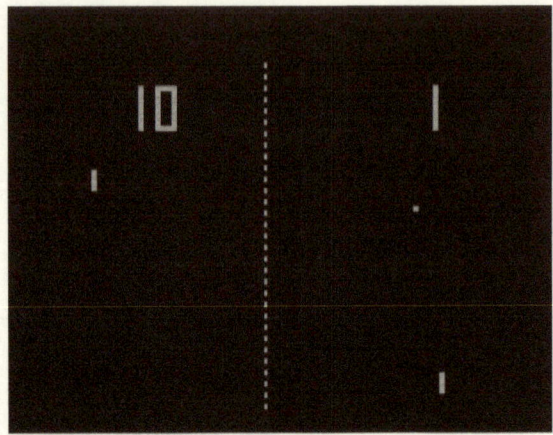

Figure 22. Pong, with a 10-1 score.

- Left: See, I was right after all! (see p. 8) I feel much better now that I understand the essence of my existence, and know why I am here.

- Right: I must say it sounds too complicated for me, and I don't understand it all. You may be right, but it doesn't change the way I play our game in the slightest.

- Left: I see that! It doesn't change the way you play, but it certainly does the way I do. Look at our score!

Conclusion (by Julia Set)

As we claimed we would do at the beginning of this book, we showed that it is now possible for anyone to formulate a scientifically reasonable theory of everything including physics and most other aspects of existence. Admittedly, we did not run any new experiments and did not produce any complex equations. Our theory is not even a GUT (Grand Unified Theory of particle physics) and it ignores several important questions in physics and cosmology, but it confidently spreads out into deeper issues such as the essence of our existence, happiness, scientific morality, religion and world politics. At least, it satisfies our personal needs. Mentally, the theory forms an interlocking structure that we had difficulty teasing apart into distinct chapters, several of which will certainly require significant corrections and additions. Undoubtedly, better writers will succeed in presenting those ideas more clearly for easier understanding, or more mathematically for deeper resolution.

Our book may only have a short life but we are convinced its ideas and concepts will flourish, whether we want it or not. Their time has come, the obvious has sprung to light and the universe is awakening!

Readers are encouraged to reject the myth of the inscrutability of modern physics and the harmful notion of non-overlapping magisteria, and to develop further their own views. The sooner we develop a scientific morality and universal governmental ethics the better. Eventually, the best views will prevail, for the greatest benefit of all humanity.

Time has come to spread the good news!

Julia Set

The last dream (by Jean Rêve)

I'm dreaming a beautiful dream. In that dream, it no longer matters that I have cancer. My spouse, my children and my friends love me, and nothing else matters. I love them too, infinitely and timelessly, and I am falling, falling, into an infinite spiral where I will blend into the Universal Consciousness. I now understand everything I needed to understand and I know that this book is helping them and all those who read it.

Jean Rêve

Figure 23. A spiralling descent towards an inevitable vanishing point. Part of a Julia set produced by a lambda sine function. On a computer screen, changing all colours in a rapid, random sequence creates a dazzling animation.

Summary

There are now enough data in physics and neuropsychology to allow the formulation of a theory of everything. The mental block
that forged a wide gap between quantum physics and the theory of
relativity as well as between science and religion can finally be lifted.

In order to understand the world we live in, we compare the odds
of various hypotheses and assign to them the relative support they
deserve. We start by showing that infinity does not exist in the material world of daily experience, and that timelessness is at the core
of quantum weirdness and virtuality. This leads to the realisation
that there are three modes of existence: concrete, virtual and abstract, and that concrete existence is the logical and inevitable consequence of abstract and virtual existence. Creation is scientifically
inevitable.

The nature of abstractness as a form of energy implies the reality of
a Universal Consciousness that diffuses into the quantum realm
and extends to every bit of matter, inert or alive. Consciousness
thus pervades the universe and constitutes the essence of our existence.

In our concrete and finite world where space, time, speed and randomness are relative, we show how we have free will to evolve
within a deterministic order. Islands of consciousness are relentlessly increasing throughout the universe, with intelligence necessarily abundant everywhere in the Cosmos.

We conclude that we are a simple link in a long evolution towards
increased consciousness, and that there are efficient and logical
ways to fulfil this goal in harmony with our nature and with each
other. We claim that all religions and human cultures desire to
reach this logical goal, and that a Universal Code of ethics based on
a scientific morality will help us settle our differences in a logical

and universally acceptable way, thereby rejecting both intolerance and the tolerance of intolerance.

We end up with a *theory of everything* that leaves us with an explanation of the weirdness of quantum physics, a sense of wonder and profound intellectual fulfillment, a tool to live happy, ethical and productive lives, and logical solutions to the world's political and existential problems.

Time has come to spread the good news!

Figures

1. Pong, the first commercially successful video game. http://en.wikipedia.org/wiki/File:Pong.png [Available: November 27, 2011]. This file is in the public domain.

2. The Wrath of Achilles, by François-Léon Benouville (1821–1859) http://en.wikipedia.org/wiki/File:Leon_Benouville_The_Wrath_of_Achilles.jpg [Available: November 27, 2011.] This photographic reproduction of a work of art is considered to be in the public domain

3. Lady Justice. Photo UKJUSTICE.jpg by T.R. Wolf. http://en.wikipedia.org/wiki/Image:UKJUSTICE.jpg [Available: October 20, 2007.] The owner released this photograph into the public domain.

4. Part of the Mandelbrot set defined by the equation $z = z^2 + c$.

5. The white Queen and Jean, the female member of a visible minority.

6. Game over!

7. A treble bull.

8. A *"nousson"*, the quantum unit of consciousness.

9. Set of three.

10. Drawing, photograph or origami?

11. Newton's method to find the cubic root of 1 in the complex plane.

12. A conscious question: "Do people really mean we are not conscious?"

13. Randomness and selection

14. Fractal: waking up and standing up.

15. Aristarchus of Samos. Statue photographed by BjørnN
     http://en.wikipedia.org/wiki/Image:Aristarchos_Samos.png
     [Available 12 December 2007.] The image was released into
     the public domain by its author.

16. Mary Wollstonecraft, painted by John Opie
     http://en.wikipedia.org/wiki/Image:Marywollstonecraft.jpg
     [Available 12 December 2007.] The image is part of the pub-
     lic domain.

17. Fire ants: 250px-Fire_ants02.jpg by Scott Bauer
     http://fr.wikipedia.org/wiki/Image:Fire_ants02.jpg [Availa-
     ble 20 Oct 2007]. The image is part of the public domain

18. From dark chaos to manifest order and universal peace.

19. March 4th celebrations.

20. Burning torch fractal.

21. Waves on the lake.

22. Pong, with a 10-1 score. Adapted from Figure 1.

23. A Julia set with an infinite spiral.

References

1. **Klevgard, Paul.** Einstein's method. [Online] [Cited: November 20, 2011.] http://www.einsteinsmethod.com/.

2. **Frank, Adam.** 3 theories that might blow up the big bang. *Discover Magazine.* [Online] April 2008. [Cited: November 20, 2011.] http://discovermagazine.com/2008/apr/25-3-theories-that-might-blow-up-the-big-bang.

3. **Heath, Sir Thomas.** *A History of Greek Mathematics.* Oxford and New York : Clarendon Press and Dover Publications, 1921, 1981. p. 446. Vol. I. From Thales to Euclid.

4. **Silagadze, Z.K.** Zeno meets modern science. *Cornell University Library.* [Online] 2005. [Cited: October 20, 2007.] http://arxiv.org/abs/physics/0505042 [physics.pop-ph].

5. **Moyer, Michael.** Is space digital? *Scientific American.* February 2012, Vol. 306, 2, pp. 31-37.

6. **Côté, G.B.** Odds in genetic counsellling. *Journal of Medical Genetics.* 1982, Vol. 19, pp. 455-457.

7. **Edwards, A.W.F.** *Likelihood.* Expanded Edition. Baltimore, London : The Johns Hopkins University Press, 1972, 1992. p. 275.

8. **Hammond, Richard.** *The unknown universe. The origin of the universe, quantum gravity, wormholes, and other things science still can't explain.* Franklin Lakes : New Page Books, 2008. p. 271.

9. **Derman, Emanuel.** Unruly humans vs the lust for order. *New Scientist.* October 11, 2011, 2835, pp. 32-33.

10. **Grousson, Mathieu.** D'où viennent les maths? *Science & Vie.* September 2007, 1080, pp. 51-67.

11. **Butterworth, B., Reeve, R., Reynolds, F., and Lloyd, D.** Numerical thought with and without words: Evidence from indigenous Australian children. *Proceedings of the National Academy of Sciences of the United States of America.* September 2, 2008, Vol. 105, 35, pp. 13179-13184.

12. **Clegg, Brian.** *A brief history of infinity. The quest to think the unthinkable.* London, U.K. : Robinson, 2003. p. 255.

13. **Dunham, William.** *Journey through genius. The great theorems of mathematics.* New York, London, Victoria, Markham, Auckland : Wiley (1990), Penguin books (1991). p. 300.

14. **Callender, C.A.** What makes time special. *FQXi Community.* [Online] 2008. [Cited: November 27, 2011.] http://fqxi.org/data/essay-contest-files/Callender_FQX.pdf?phpMyAdmin=0c371ccdae9b5ff3071bae8 14fb4f9e9.

15. —. Is time an illusion? *Scientific American.* 2010, Vol. 302, 6, pp. 59-65.

16. **Musser, G.** Where do space and time come from? New theory offers answers, if only physicists can figure it out. [Online] April 12, 2012. [Cited: April 19, 2012.] http://www.scientificamerican.com/observations/2012/04/12/where-do-space-and-time-come-from-new-theory-offers-answers-if-only-physicists-can-figure-it-out/.

17. **Tributsch, H.** Quantum paradoxes, time, and derivation of thermodynamic law: opportunities from change of energy paradigm. *Journal for General Philosophy of Science.* 2006, Vol. 37, pp. 287-306.

18. **Falk, Dan.** *In Search of Time. Journeys along a Curious Dimension.* Toronto : Emblem. McClelland & Stuart., 2008. p. 330.

19. **Folger, Tim.** Time may not exist. *Discover Magazine.* [Online] June 12, 2007. [Cited: November 27, 2011.] http://discovermagazine.com/2007/jun/in-no-time/.

20. **Lockwood, Michael.** *The labyrinth of time. Introducing the universe.* Oxford : Oxford University Press, 2005. p. 405.

21. **Barbour, Julian.** *Platonia.* [Online] [Cited: November 27, 2011.] http://platonia.com.

22. **Gilder, Louisa.** *The Age of Entanglement. When quantum physics was reborn.* New York : Knopf, 2008. p. 446.

23. **Einstein A., Podolsky B., Rosen, N.** Can quantum-mechanical description of physical reality be considered complete? *Physical Review.* 1935, Vol. 47, pp. 777-780.

24. **Bell, J.S.** On the Einstein Podolsky Rosen paradox. *Physics.* 1964, Vol. 1, pp. 195-200.

25. **Aspect A., Grangier, P., Roger, G.** Experimental realization of Einstein-Podolsky-Rosen-Bohm Gedankenexperiment: A new violation of Bell's inequalities. *Physical Review Letters.* 1982, Vol. 49, 2, pp. 91-94.

26. **Pajot, Philippe.** L'expérience ultime qui a défié le photon. *Science & Vie.* September 2007, 1080, pp. 68-72.

27. **Salart, D., Baas, A., Branciard, C., Gisin, N., Zbinden, H.** Testing the speed of 'spooky action at a distance'. *Nature.* 2008, Vol. 454, pp. 861-864.

28. **O'Connell, A.D., Hofheinz, M., Ansmann, M., Bialczak, R.C., Lenander, M., Lucero, E., Neeley, M., Sank, D., Wang, H., Weides, M., Wenner, J., Martinis, J.M., Cleland, A.N.** Quantum ground state and single-phonon control of a mechanical resonator. *Nature.* 2010, Vol. 464, pp. 697-703.

29. **Barone, M. and Gajewska, M.** Einstein and the mystery of eternity of life. *Edukacaja humanistyczna.* [Online] 2009, Vol. 1, 16. [Cited: November 27, 2011.] http://arxiv.org/ftp/arxiv/papers/0904/0904.1280.pdf.

30. **Gribbin, John.** *In Search of Schrödinger's Cat. Quantum Physics and Reality.* Toronto, New York, London, Sydney, Auckland : Bantam Books, 1984. p. 303.

31. **Heisenberg, W.** Über den anschaulichen Inhalt der quantentheoretischen Kinematik und Mechanik. 1927, Vol. 43, (S), pp. 172-198.

32. **Kurzweil, Ray.** *The Singularity is Near. When humans transcend biology.* New York : Viking Penguin, 2005. p. 652.

33. **Gauger, E., Rieper, E., Morton, J.J.L., Benjamin, S.C., Vedral, V.** Quantum coherence and entanglement in the avian compass. *www.arxiv.org.* [Online] 2009. [Cited: December 3, 2011.]

http://arxiv.org/PS_cache/arxiv/pdf/0906/0906.3725v3.pdf [quant-ph].

34. **Engel, G.S., Calhoun, T.R., Read, E.L., Ahn, T.-K., Mančal, T., Cheng, Y.-C., Blankenship, R.E., Fleming, G.R.** Evidence for wavelike energy transfer through quantum coherence in photosynthetic systems. *Nature.* 2007, Vol. 446, pp. 782-786.

35. **Collini, E., Wong, C.Y., Wilk, K.E., Curmi, P.M.G., Brumer, P., Scholes, G.D.** Coherently wired light-harvesting in photosynthetic marine algae at ambient temperature. *Nature.* 2010, Vol. 463, pp. 644-647.

36. **Rieper, E., Anders, J., Vedral, V.** The relevance of continuous variable entanglement in DNA. *www.arxiv.com.* [Online] June 21, 2010. [Cited: December 3, 2011.] http://arxiv.org/PS_cache/arxiv/pdf/1006/1006.4053v1.pdf [quant-ph].

37. **Hameroff, S.** Overview: Could life and consciousness be related to the fundamental quantum nature of the Universe? *Quantum Consciousness.* [Online] [Cited: October 30, 2010.] http://www.quantumconsciousness.org/overview.html.

38. **Penrose, Roger.** *The emperor's new mind. Concerning computers, minds, and the laws of physics.* Oxford : Oxford University Press, 1989. p. 602.

39. **Seife, Charles.** *Decoding the Universe. How the science of Information is Explaining Everything in the Cosmos, from our Brains to Black Holes.* New York, Toronto, London : Penguin Group, 2006. p. 296.

40. **Casimir, H.B.G.** On the attraction between two perfectly conducting plates. *Proc. Kon. Nederland. Akad. Wetensch.* 1948, Vol. B51, pp. 793-795.

41. **Wilson, C.M., Johansson, G., Pourkabirian, A., Simoen, M., Johansson, J.R., Duty, T., Nori, F., & Delsing, P.** Observation of the dynamical Casimir effect in a superconducting circuit. *Nature.* November 17, 2011, Vol. 479, pp. 376-379.

42. **Hawking, Stephen W.** Black Hole explosions? *Nature.* 1974, Vol. 248, 5443, pp. 30-31.

43. —. *A Brief History of Time. From the big bang to black holes.* Toronto, New York, London, Sydney, Auckland : Bantam Books, 1988. p. 198.

44. **Delahaye, Jean-Paul.** L'ensemble de tous les ensembles. *Pour la Science.* November 2010, 397, pp. 146-151.

45. **Holmes, M. Randall.** Elementary set theory with a universal set. [Online] 1998-2010. [Cited: December 3, 2011.] http://math.boisestate.edu/~holmes/holmes/head.pdf.

46. **Moss, Lawrence S.** Non-wellfounded Set Theory. *The Stanford Encyclopedia of Philosophy.* [Online] August 17, 2009. [Cited: December 3, 2011.] http://plato.stanford.edu/archives/fall2009/entries/nonwellfounded-set-theory.

47. **Hawking, S.W. and Penrose, R.** The singularities of gravitational collapase and cosmology. *Proceedings of the Royal Society of London. A.* 1970, Vol. 314, pp. 529-548.

48. **Hawking, S. and Mlodinow, L.** *The grand design.* New York : Bantam Books, 2010. p. 199.

49. **Krauss, Lawrence M.** *A universe from nothing. Why there is something rather than nothing.* s.l. : Free Press. Simon & Schuster, 2012. p. 204.

50. **Stenger, Victor J.** *The fallacy of fine-tuning. Why the universe is not designed for us.* Amherst, New York. : Prometheus Books, 2011. p. 345.

51. **Weinberg, Steven.** *The first three minutes. A modern view of the origin of the universe.* U.S.A. : Basic books. HarperCollins, 1977, 1988. p. 198.

52. **Gribbin, John.** *In search of the mutliverse.* London : Penguin Books, 2009. p. 228.

53. **Wikipedia.** Many-worlds interpretation. *Wikipedia.* [Online] [Cited: March 27, 2012.] http://en.wikipedia.org/wiki/Many-worlds_interpretation.

54. **Candelas, P., de la Ossa, X., He, Y.-H., Szendrői, B.** Triadophilia: a special corner in the landscape. *Adv. Theor. Math.*

Phys., 12(2). [Online] 2008. [Cited: December 4, 2011.] http://arxiv.org/abs/0706.3134 [hep-th].

55. **Consolmagno, Guy.** *God's mechanics. How scientists and engineers make sense of religion.* Mississauga (Ontario) : Jossey-Bass, Wiley, 2008. p. 245.

56. **Pereira, A. Jr, Edwards, J.C.W., Lehmann, D., Nunn, C., Trehub, A. & Velmans, M.** Understanding consciousness. A collaborative attempt to elucidate contemporary theories. *Journal of Consciousness.* 2010, Vol. 17, 5-6, pp. 213-219.

57. **Velmans, Max.** *Understanding Consciousness.* 2nd edition. London : Routeledge, 2009.

58. **Lambert, Frank L.** Entropy sites. A guide. [Online] [Cited: October 31, 2007.] www.entropysite.com.

59. **Vedral, Vlatko.** *Decoding Reality. The Universe as Quantum Information.* s.l. : Oxford University Press, 2010. p. 229.

60. **Wikipedia.** Claude Shannon. *Wikipedia.* [Online] April 10, 2012. [Cited: April 15, 2012.] http://en.wikipedia.org/wiki/Claude_Shannon.

61. **Gleick, James.** *The information. A history, a theory, a flood.* New York : Vintage Books, 2011, 2012. p. 527.

62. **Gödel, Kurt.** Über formal unentscheidbare Sätze der Principia Mathematica und verwandter Systeme, I. *Monatshefte für Mathematik und Physik.* 1931, Vol. 38, pp. 173-198.

63. **Goldstein, Rebecca.** *Incompleteness. The Proof and Paradox of Kurt Gödel.* New York and London : Norton & Co., 2005. p. 296.

64. **Byers, William.** *The blind spot. Science and the crisis of uncertainty.* Princeton and Oxford : Princeton University Press, 2011. p. 208.

65. **Gregory, R. and Cavanagh, P.** The Blind Spot. *Scholarpedia, 6: 9618.* [Online] 2011. [Cited: April 1, 2012.] http://www.scholarpedia.org/article/The_Blind_Spot.

66. **Pepperberg, Irene M.** *Alex & Me: How a scientist and a parrot discovered a hidden world of animal intelligence--and formed a deep bond in the process.* Toronto : HarperCollins, 2008. p. 232.

67. **Weinberg, W.** Über den Nachweis der Vererbung beim Menschen. *Jahreshefte des Vereins für vaterländische Naturkunde in Württemberg*. 1908, Vol. 64, pp. 368-382.

68. **Edwards, A.W.F.** Perspectives. Anecdotal, historical and critical commentaries on genetics. G.H. Hardy(1908) and Hardy-Weinberg equilibrium. *Genetics*. July 2008, Vol. 179, 3, pp. 1143-1150. http://www.genetics.org/content/179/3/1143.full.

69. **Gleick, James.** *Chaos. Making a new science.* New York : Penguin, 1987. p. 354.

70. **Peitgen, H.-O. and Richter, P.H.** *The beauty of fractals. Images of complex dynamical systems.* Berlin : Springer-Verlag, 1986. p. 199.

71. **Wegner, Timothy and Peterson, Mark.** *Fractal creations.* Mill Valley : Waite Group Press, 1991. p. 315.

72. **Libet, B., Gleason, C.A., Wright, E.W., Pearl, D.K.** Time of conscious intention to act in relation to onset of cerebral activity (readiness-potential). The unconscious initiation of a freely voluntary act. *Brain.* September 1983, Vol. 106, Part 3, pp. 623-642.

73. **Harris, Sam.** *Free will.* New York : Free press, 2012. p. 85.

74. **Gamble, M.** Universal Consciousness: An Historical and Scientific Perspective. *The LIGHTouch.* [Online] 1997. [Cited: September 15, 2010.] http://www.lightouch.com/conscious.htm.

75. **Budnik, Paul.** *What Is and What Will Be. Integrating Spirituality and Science.* Los Gatos : Mountain Math Software, 2006. p. 229.

76. **Bohm, D. and Hiley, B.J.** *The Undivided Universe: an Ontological Interpretation of Quantum Theory.* London : Routledge, 1993. p. 407.

77. **Blake, William.** Auguries of Innocence (1863). [book auth.] D.V. Erdman (ed). *The Complete Poetry of William Blake.* New York : Anchor Books, Random House, 1988, pp. 490-495.

78. **Dalai Lama, Bstan-'dzin-rgya-mtsho.** *The universe in a single atom: the convergence of science and spirituality.* New York : Morgan Road Books, Random House, 2005. p. 216.

79. **Dawkins, Richard.** *The selfish gene.* Oxford, New York : Oxford University Press, 1976 (2nd edition, 1989). p. 352.

80. **Barton, N.H.** Genetic hitchhiking. *Phil. Trans. R. Soc. Lond. B.* 2000, Vol. 355, pp. 1553-1562.

81. **Bryson, Bill.** *A short history of nearly everything.* Toronto : Doubleday Canada, 2003, 2005. p. 624.

82. **Teresi, Dick.** *Lost discoveries. The ancient roots of modern science - from the Babylonians to the Maya.* Riverside, N.J. : Simon & Schuster paperback, 2002. p. 453.

83. **Sagan, Carl.** *Cosmos.* s.l. : Ballantine Books Edition, 12th printing, 1990, 1980. p. 325.

84. **Schultz, Ted R.** In search of ant ancestors. *Proceedings of the National Academy of Sciences.* December 19, 2000, Vol. 97, 26, pp. 14028-14029.

85. **Keller, Laurent and Gordon, Élisabeth.** *The lives of ants (translated from the French by James Grieve).* Oxford : Oxford Universtiy Press, 2009. p. 252.

86. **Sykes, Bryan.** *Adam's curse.* New York : Norton, 2003, 2004. p. 320.

87. **Teilhard de Chardin, Pierre.** *The phénomène humain.* Paris : Les éditions du Seuil, 1955. p. 347.

88. **Dawkins, Richard.** *The God Delusion.* New York : Houghton Mifflin, 2006. p. 406.

89. **Volta, Ornella.** *Guida dell'altro mondo.* Milan : Editoriale Milanese, 1970. p. 341.

90. **Conesa, Pierre.** *Guide du Paradis.* Paris : Éditions de l'Aube., 2004. p. 173.

91. **Henderson, Bobby.** *Church of the Flying Spaghetti Monster.* [Online] http://www.venganza.org/.

92. **National Center for Biotechnology Information.** Microarrays: Chipping away at the mysteries of science and medicine. *A Science Primer.* [Online] July 27, 2007. [Cited: December 26, 2011.] http://www.ncbi.nlm.nih.gov/About/primer/microarrays.html.

93. **The international cell death society.** [Online] [Cited: December 31, 2011.] http://www.celldeath-apoptosis.org/.

94. **Lobo, I.** Genomic imprinting and patterns of disease inheritance. *Scitable, Nature Educaton.* [Online] 1(1),2008. [Cited: December 26, 2011.] http://www.nature.com/scitable/topicpage/genomic-imprinting-and-patterns-of-disease-inheritance-899.

95. **Kent, W.J., Sugnet, C.W., Furey, T.S., Roskin, K.M., Pringle, T.H., Zahler, A.M., Haussler, D.** The human genome browser at UCSC. *Genome Research.* June 12, 2002, Vol. 12, 6, pp. 996-1006.

96. **UCSC Genome Bioinformatics Group.** *UCSC Genome Browser.* [Online] [Cited: August 30, 2009.] http://genome.ucsc.edu/.

97. **Kurzweil, Ray.** *The age of spiritual machines. When computers exceed human intelligence.* New York : Viking Penguin, 1999. p. 388.

98. **St-Exupéry, Antoine de.** *The little prince (Translated from the French "Le petit prince" by Katherine Woods, 1945).* London : Pan Books, 1945.

99. **Pruche, Benoît.** *Histoire de l'homme, mystère de Dieu. Une théologie pour les laïcs.* Montréal : Éditions du lévrier, Desclée de Brouwer, 1961. p. 452.

100. **Gould, Stephen Jay.** http://www.stephenjaygould.org/library/gould_noma.html. *from: Gould, S.J., Nonoverlapping Magisteria, Natural History 106: 16-22, March 1997.* [Online] [Cited: December 28, 2011.]

101. **Harris, Sam.** *The moral landscape.* New York : Free Press. Simon & Schuster., 2010. p. 308.

102. **Deutsch, David.** *The beginning of infinity. Explanations that transform the world.* London : Allen Lane, 2011. p. 487.

103. **Spinney, L.** The underhand ape. *New Scientist.* November 5, 2011, Vol. 212, 2837, pp. 42-45.

104. **King, Sir David.** Rigour, respect and responsibility: a universal ethical code for scientists. [Online] 2007. [Cited: December 29, 2011.] http://www.bis.gov.uk/assets/BISPartners/GoScience/Docs/U/universal-ethical-code-scientists.pdf.

105. **Payne, D. and Landry Brett, J.L.** A uniform code of ethics: business and IT professional ethics. *Communications of the ACM.* 2006, Vol. 49, 11, pp. 81-84.

106. **Klein, Naomi.** *No logo: taking aim at the brand bullies.* Toronto : Vintage Canada Editions, 2000. p. 490.

107. —. *Naomi Klein.* [Online] [Cited: March 4, 2008.] http://www.naomiklein.org.

108. **Robinson, B. A.** Stages of development, from an ovum & spermatozoon to a newborn. *Religious tolerance.* [Online] November 27, 2011. [Cited: December 29, 2011.] http://www.religioustolerance.org/abo_fetu.htm.

Index

About the authors

The authors are Canadians without international stature, a regular scientist and a part-time philosopher. Their names are ambiguous pseudonyms meant to prevent the categorisation or rejection of their ideas based on age, cultural background, gender or pigmentation. The first names Julia and Jean are assumed to belong to women in English, while Jean is assumed to be a man in French; both Julia and Jean can be surnames; a Julia set is a mathematical object discovered by Gaston Julia; Set or Seth is the ancient Egyptian god of chaos; and Jean Rêve (or *"j'en rêve"*) is an easy pun meaning "I dream of it" in French. In case of success or discredit, the authors' anonymity will safely protect their modesty or their local reputation.

The scientific chapters were mainly written by Julia, and the fanciful dialogues by Jean, with a good deal of mutual interference.

As this book is going to press, the authors can be reached by email at the following addresses:

juliaset@live.ca and jeanreve@hotmail.ca.

Eventually, this will no longer be true.

www.ingramcontent.com/pod-product-compliance
Lightning Source LLC
Chambersburg PA
CBHW032013170526
45157CB00002B/677